T0298424

Emotion in the Design Process

Emotion in the Design Process investigates and demonstrates the tight connections between emotion and the design process. Manipulating the design process can be a stressful learning experience and some students have difficulty deciding how to resolve their design problems. This book explores and illustrates the close relationship between emotion and the design process by using new models and perspectives under the umbrella of 'design and emotion'.

This title reveals that a greater understanding of design and emotion can inspire design students to explore how emotion can affect their decision-making and design processes. It enables the reader to develop methods to control their emotions to make effective decisions and strengthen their ability to manipulate the design process. This book features a study that develops a design process model in order to make the decision-making processes more transparent. With a focus on the investigation of the 'intrinsic factors', this book features quantitative and qualitative research methods. Underpinned by deep-level research, the book outlines the strengths and limitations of the study and reveals the findings to create decision-making models where emotion is considered. Case studies are included to show the theories in practice.

By reading this book, design students, who are often confused by the design process, will be able to grasp it and learn to regulate their emotions as a result while also producing better designers that can improve the overall quality and standard of the design industry. As such, this book will appeal mostly to students, researchers, and academics in any field where design is a key task. It will also fascinate anyone who is interested in Design and Emotion, Kansei Design and Engineering, and Design and Technology.

Amic G. Ho is a design scholar, typographer, and communication designer. He is Assistant Professor and Programme Leader of BFA (Hons) in Imaging Design & Digital Art and BFA (Hons) in Cinematic Design & Photographic Digital Art in the Department of Creative Arts at Hong Kong Metropolitan University. His design expertise and research interests are in typography, emotion and design, and communication design.

Design, Emotion and Creativity
Series Editor: Amic G. Ho

This book series explores the close relationships between design, emotional experiences, and the creative process. The books presented here comprehensively understand how these three aspects intertwine and influence one another while offering theories, insights, and strategies for designers, researchers, and creative students alike. Themes within this series include the role of emotion in design, investigations into the creative process, innovative design methods and techniques, design thinking, user experience (UX), computing and emotion recognition, and the future trends and challenges for emotion in the design process. Aimed at those who aim to design more creatively and innovatively, please get in touch to discuss your book idea for this series.

Emotion in the Design Process
Intrinsic Factors on Emotion Management for Decision-making
Amic G. Ho

Emotion in the Design Process

Intrinsic Factors on Emotion Management for Decision-making

Amic G. Ho

CRC Press
Taylor & Francis Group
Boca Raton London New York

CRC Press is an imprint of the
Taylor & Francis Group, an **informa** business

Designed cover image: Amic G. Ho

First edition published 2024
by CRC Press
2385 NW Executive Center Dr. Suite 320, Boca Raton, FL 33431

and by CRC Press
4 Park Square, Milton Park, Abingdon, Oxon, OX14 4RN

CRC Press is an imprint of Taylor & Francis Group, LLC

© 2024 Amic G. Ho

ISBN: 9781032484112 (hbk)
ISBN: 9781032468082 (pbk)
ISBN: 9781003388920 (ebk)

DOI: 10.1201/9781003388920

Typeset in Times
by Newgen Publishing UK

Contents

Acknowledgements

After having worked on my PhD a decade ago, the publication of my first book marks another very important moment for me.

Now that my PhD is complete, I would like to extend special thanks to those people who supported me during my studies. In particular, I would like to thank my supervisor Professor Siu K. W. Michael for not only teaching me but also enlightening me and for his insightful comments as my research progressed.

Special thanks are also due to Ms Ruth Chau who graciously lent her time in supporting me in this study, and to all my students who supported me with their joy and love as I progressed.

My thanks also greatly extend to the professors, researchers, and experts who raised the topic of 'design and emotion' to mark an important record in human knowledge for new aspects of information in the field.

Finally, and most importantly, I would like to sincerely thank my mother, who not only brought me into this world but always supported me as I made my contributions to design research.

The topic covered in this book aims to explore and address the possibilities regarding the application of design and emotion concerns to design processes and studies. Thanks are due to those experts, professors, and researchers who made an enormous contribution to the field over the decades. It is believed that there are always new opportunities in this field to further develop and sustain enquiry in design research. The author always welcomes any constructive discussion from any professionals to advise and enrich this topic and field.

Figures

1 Introduction

INTRODUCTION

Throughout the last few decades, numerous scholars have concentrated on the design process as a whole. They have examined how to govern the whole design process (Best, 2006) or how to run the design more effectively via the use of alternative methodologies (Noble and Bestley, 2005; Peto, 1999). Several of them went so far as to establish design procedures (Chen and Chen, 2004; Cross and Sivaloganathan, 2004) for certain accomplishments. Conversely, few of them concentrated on assisting inexperienced design students/designers in operating their design processes at a basic level and teaching them how to commence projects quickly and efficiently.

Many inexperienced design students may be overjoyed to gain acceptance to design school and eager to learn more. They may be inquisitive about the content of design studies and enquire regarding how to create something worthwhile. However, in most situations, inexperienced design students/designers at the entry level confront challenges when they begin their design assignments following their instructors' briefings (Gibbons, 2003; Marshall and Rowland, 1993). Inexperienced design students/designers at the entry level may be unsure of how to begin or may have difficulty implementing various aspects of the design process, including defining the design topic, forming design concepts, conducting appropriate research, planning the design process, managing time, embodying the design, and performing detailed design. Typically, this development process is comprised of a series of subproblems that must be solved, with each step requiring the completion of multiple problem-solving activities (Pahl and Beitz, 1996). While instructors frequently give valuable suggestions for managing their design processes step by step, not all students are able to apply them smoothly and successfully. Some students eventually quit the design project, becoming disoriented by the design directions or needing clarification on the lengthy design-development process.

Additionally, their lack of design expertise impacts their personal capacity to make appropriate judgements during the design process. Various techniques have been created in past research, in conjunction with instructions from school tutors, to guide designers and students towards good design practice (Govers, Hekkert, and Schoormans, 2004; Mann, 2004). This investigation demonstrates that some design teachers seek to increase starting design students' ability to make sound judgements,

DOI: 10.1201/9781003388920-1

find the essential information to complete a project, and identify when they need help. Others construct criteria for the efficient administration of the design process, which include self-motivation, time management, self-evaluation, and the capacity to combine these talents to generate excellent work on time. However, it is still true that inexperienced design students/designers at the entry level are unable to successfully manage their design processes. Several previous studies have investigated different approaches to directing novice design students towards effective practice (Govers, Hekkert, and Schoormans, 2004; Mann, 2004). This particular line of inquiry elucidates that certain design educators endeavour to enhance the aptitude of novice design students in terms of making sound judgements, acquiring the requisite knowledge to complete a task, and discerning when they require aid. Various scholars have formulated principles for proficiently managing the design process, encompassing self-initiative, temporal organisation, self-evaluation, and the capacity to integrate these proficiencies to generate high-calibre output within the designated timeframe.

Notwithstanding these endeavours, a considerable number of novice design students remain incapable of commencing their design projects proficiently due to their lack of familiarity with the stages of the design process and their inability to adjust to the techniques. Furthermore, there could be additional hindrances that obstruct the overall design process. This implies that the students may not completely comprehend or adapt to these strategies. Are there any more hidden issues impeding their design processes in general? Are the issues connected to their thought habits or perhaps their creative processes? The aforementioned variables could potentially exhibit a correlation with the cognitive patterns or creative ideation of the students. To examine and comprehend this problem, it is necessary to do a thorough review of the literature on students' innate responses to the design process.

1.2 IDENTIFICATION OF THE RESEARCH PROBLEM

1.2.1 WHAT IS THE DESIGN PROCESS?

Prior to delving into the factors contributing to the challenges faced by inexperienced learners in comprehending and implementing established design methodologies, it is imperative to conduct an inquiry into the concept of 'comprehensive design procedure'. Over the course of recent decades, a multitude of definitions pertaining to the design process have been formulated. During the 1960s, several noteworthy studies were conducted on design processes, including the analysis–synthesis model (Archer, 1965) and the systematic design approach (Jones, 1984). The analysis–synthesis model represents the initial model developed to elucidate the process by which inexperienced students engage in problem analysis, comprehensively define the problem, and subsequently synthesise a solution. The initial inquiries primarily centred on discerning a cognitive and logical foundation for the process of design. Jones (1984) introduced and expounded upon the cognitive rational methodology. The individual may utilise both intuitive experience and rigorous logical argumentation to advance Archer's analysis–synthesis model into a fundamental concept of systematic design. The concept of systematic design elucidates the manner in which designers employ their intuitive expertise and rational thought processes to generate concepts and resolutions. The notion has gained significant traction among design scholars and educators as a potent pedagogical strategy for neophyte design learners. This initial iteration of the design process, akin to contemporaneous models, is underpinned by a robust rational framework. To understand and implement effective strategies for their design processes, students and novice designers must first understand some fundamental aspects of the design process. There are several definitions of the design process, all of which were exhaustively developed in the 2000s.

The UK charity Design Council examined numerous successful projects and documented their design processes through all stages to serve as references for creative teams and experts involved in board-design disciplines. It defined the five primary stages that designers must progress through. The model, which Visocky O'Grady and Visocky O'Grady cited in *A Designer's Research Manual* (2017, p. 104), divides the design process into the five functional stages depicted in the model. The Design Council rated this model by J. Visocky O'Grady and K. Visocky O'Grady as the most functional (2017).

Additionally, the design process (Best, 2006) comprises a number of approaches that are incorporated to meet the particular requirements of individual projects and is a way of formalising and adapting the design to satisfy the demands of customers. In *Design Management: Managing Design Strategy, Process and Implementation*, according to Best (2006, p. 117), the process of design comprises a range of techniques that are tailored to meet the specific needs of various projects. He also emphasised the inherent character of the design process:

Design processes are not linear, as there are many feedback loops built in to allow for the iterative nature of design and to accommodate the insights gained at each stage of the process.

They are difficult to standardise, in part because of their iterative, non-linear nature and also because the needs of clients and users are so different. (Best, 2006, p. 26).

Additionally, many critical steps within the design process were addressed. Apart from the Design Council's model, Best (2006) proposed that it is comprised of six major stages, each with distinct activities and achievable outputs. Simultaneously, the publication featured Rollestone's recommended design methodology (2003) in his paper, *Scenario-Based, Value-Driven Design Methods*. This provided an overview of the entire web-based design process, which is represented clearly.

Foque (1995) presented an alternative design process paradigm, beginning with 'problem formulation' and concluding with 'realisation'. On one hand, the standard design process contains a summary of important phases, which include problem definition, problem analysis, concept development, assessment, decision, and realisation (Noble and Bestley, 2005). On the other hand, the book *Visual Research* discussed a significant case study in which graphic designer Matt Cooke established a pragmatic approach to design methodology throughout the design process (Noble and Bestley, 2005). According to his proposal, the entire process should be divided into four distinct stages: definition, divergence, transformation, and convergence.

According to Karl (2006), the design process encompassed all stages of a project as well as the time it took to deliver the completed result. An identification stage was followed, in order, by a conception stage, an exploration stage, a definition/modelling stage, and lastly a production stage. Moreover, it necessitated a significant amount of decision-making throughout the whole process. On pages 162–172 of their paper, Brissaud, Garro, and Poveda (2003) unequivocally emphasised that decision-making during the design process resulted in the final design solution being implemented. Brissaud et al. (2003, p. 164) also observed that:

> progress in the decision-making process assumed to be representative of an actual design process.

It was noted that the ultimate design solution is a result of the decision-making process. Thus, it can be asserted that the process of decision-making exerts a significant impact on the eventual results of design. The interplay between the design process and its outcome is such that the manipulation of the former can significantly impact the latter. As a result, decision-making and outcome evaluation are intrinsically linked. Therefore, it is imperative to establish a design process model that can elucidate the decision-making process. The advancement of the decision-making process is regarded to be representative of the progress of a real-world design process. It is reasonable to assert that the process of decision-making has a significant impact on the ultimate conclusion. In contrast, the process reflected in the design solution is the consequence of a decision that has already been taken.

As a result, decision-making and outcome evaluation are tightly connected processes. It is, therefore, necessary to develop a design process model to make the decision-making processes more transparent. It is impossible to standardise the design process since it is not linear but iterative in nature and varies according to the particular requirements of each project and its clients. The design process comprises beginning and finishing projects. Also, there are various steps or phases that have to be fulfilled to do the required tasks. The decision-making process is divided into many stages that must be completed before suitable judgements may be formed.

1.2.2 THE DESIGN PROCESS AND DECISION-MAKING

As we can see from the previous explanation, the design process is a methodical approach to investigating and developing particular design solutions that follow a set of rules. There is a great deal of decision-making involved in the design process (Levin, 1984). In addition, design projects frequently entail a high degree of ambiguity, given their reliance on objectives that lack precise delineation and entail intricate organisational structures. Designers are faced with the challenge of working efficiently amidst this difficulty. To address this, they must generate a variety of plans and employ their judgement in selecting from various parameters. During the design process, designers strategise, arrange, and frequently oversee their design colleagues in carrying out their determinations (Levin, 1984). Designers must make judgements on a wide range of issues, including solution formulation, consistency testing, and comparison and selection of alternatives (Herrmann and Schmidt, 2002). Effective decision-making contributes significantly to the prescription and optimisation of the quality of the solution to the issue. To rephrase, the ability to make suitable judgements is essential to enhance design outputs and achieve success. In order to comprehend how individuals think, choose, and make decisions, it is necessary to grasp their thought processes.

According to Almendra and Christiaans' (2009) research, the process of decision-making is contingent upon factors such as the ability to access and manage knowledge, effective communication skills, and the utilisation of a strategic plan to resolve problems. The inclusion of these factors has an impact on the decision-making process and ultimately affects the outcome of the design process.

Furthermore, it has been determined that enhancing the managerial aptitude of designers is a crucial component of the design process, which also encompasses a significant amount of decision-making (Scaletsky and Marques, 2009). Scaletsky and Marques (2009) posited that designers ought to acquire materials-allocation skills, given that the selection of materials is a critical determinant of project success. Furthermore, the authors suggested that the designer ought to consider factors such as cost, aesthetics, and material strength during the design process. In general, proficient decision-making has the potential to enhance the organisation of the design process by facilitating task planning, conflict resolution, and the resolution of difficulties or low motivation in the design process. This study aims to investigate the approach of novice design students towards the design process: specifically, their decision-making process and cognitive strategies employed while completing a project. Following an initial assessment of the existing literature, potential strategies for enhancing students' decision-making abilities throughout all stages of the design process will be explored.

1.2.3 Emotion and Decision-making

Scholars have carried out crucial studies to better comprehend decision-making processes and procedures. Antonio Damasio, a neurologist, performed a study on individuals who had been damaged in a specific area of their brains, the ventromedial prefrontal cortex. The results were published in the journal Neuron (vmPFC). Despite the fact that the patients' emotional systems were damaged, their cognition and problem-solving abilities were generally normal. Due to an inability to learn from prior mistakes, these patients participated in a pattern of decision-making that resulted in unfavourable outcomes on several occasions (i.e. they could not make proper choices in the real world). As a result of poor vmPFC function, these patients were unable to make decisions or perform their jobs efficiently. Even though they could define what they should accomplish and how they should operate, they were unable to make clear choices on how to go about doing so effectively (Damasio, 1994). As a result, Damasio established the somatic-marker theory that says emotion plays a significant role in decision-making, particularly in instances where someone's choice entails unclear consequences with either reward or punishment as a possible conclusion (Damasio, 1994; Naqvi, Shiv, and Bechara, 2006). This theory was subsequently investigated to demonstrate its validity (Bechara and Damasio, 2005; Naqvi, Shiv, and Bechara, 2006), and it was further supported by a series of similar studies conducted by other neuroscientists (Bechara et al., 2005).

Several alternative hypotheses have been posited to elucidate the role of affect in the process of making choices. Desmet's (2008) basic emotion theory posited that emotions are the result of the interplay between mood and behaviour. Desmet's study on photo journalling demonstrated that designers' design process methods can be influenced by various emotions, resulting in diverse outcomes that elicit distinct emotional responses. Desmet postulated that the incorporation of positive emotions is advantageous in the field of design, based on his interactions with the designers in the photo journal study. The majority of designers exhibited a tendency to sustain a positive emotional state during the design process, as they perceived it to be advantageous for producing favourable results. Desmet's theoretical framework regarding the impact of positive emotions aligns with the findings of a scholarly investigation conducted by Isen (1993), who suggested that positive emotions yield favourable outcomes in the process of decision-making. Isen's research revealed that individuals who experienced positive emotions demonstrated superior abilities in making judgements, ratings, categorisations, and other related tasks, compared to those who maintained a neutral emotional state. As per his proposition, emotions hold significant importance in the process of design and creativity.

These outcomes illustrate how improving people's emotional or affective systems might help them make quicker judgements between good and poor options while also reducing cognitive overload. Norman (2004) also claimed that emotional systems were intimately related to behaviour and that they assisted the body in preparing itself to respond appropriately in specific situations. According to the outcomes of this research, the common understanding of the effect of emotions on decision-making acknowledges that they help individuals to make acceptable judgements when presented with a variety of alternatives.

The findings of these investigations suggest that decision-making is not solely a product of rational and logical cognition. Rather, enhanced emotional or affective mechanisms can facilitate individuals in making prompt decisions between favourable and unfavourable options, thereby mitigating cognitive overload. Unsurprisingly, research has demonstrated a strong correlation between emotions and behaviour as well as their role in facilitating physiological readiness for contextually appropriate reactions (Norman, 2004). According to the existing literature, emotions are believed to facilitate the making of suitable choices when individuals are presented with various alternatives. For design studies, we know that decision-making is a necessary part of the process of creating something new. What role do emotions play in the design process, and how do they influence decision-making throughout the whole process? It remains to be seen whether there is a link between positive emotions and a successful design process that leads to a successful design output.

1.2.4 EMOTION AND DESIGN PROCESS

The design process is characterised by the need for decision-making. The potential influence of designers' emotions on their decision-making capacity has been a subject of inquiry among scholars. Research has indicated that the experience of positive emotions may enhance cognitive processing abilities during the design phase

(Kaufmann, 2003). The study revealed that the designers exhibited superior analytical skills and were able to make informed decisions based on the information at hand. As information analysis, also known as information processing, constitutes a form of decision-making, it seems that favourable emotions have the potential to augment the comprehensive design procedure.

The comprehension of potential users' emotions can impact the decision-making process in design, in addition to the influence of designers' own emotions. Several research studies have investigated the process by which designers acquire knowledge about the thoughts and emotional states of others (Koskinen, Battarbee, and Mattelmäki, 2003). According to Koskinen, Battarbee, and Mattelmäki (2003), certain research methodologies, such as the observation of users' design experiences and the collection of users' visual cues, can facilitate the augmentation of designers' empathic comprehension. The desire to comprehend the thoughts and emotions of others is a fundamental human sentiment that fosters intimate interpersonal connections. The study investigated the correlation between designers' comprehension of consumers' requirements and their ideation processes. Through their study, Salovey and Lopes (2004) revealed that the ability to comprehend the thoughts and emotions of others was a necessary component in the observation of users. The utilisation of empathic senses, such as customer observation; data collection through visual, auditory, and sensory cues; data analysis; and brainstorming, were employed during the process of data collection and analysis. The authors noted that incorporating empathic considerations throughout the design process facilitated the identification of potential consumer needs and innovative ideas. They emphasised the significance of designers' capacity to comprehend the perspectives of others.

Emotions have the potential to exert an influence on decision-making during the entirety of the design process as well as on the perception of others' needs. This inquiry pertains to the extent to which various emotions exert comparable effects on decision-making during distinct phases of the design process and whether any associations exist between emotion and efficacious decision-making that engender more effective design procedures. As previously stated, despite the implementation of diverse cognitive and rational techniques aimed at initiating neophyte design students into the design process, a considerable number of students encounter difficulties in proficiently manipulating their designs. The incorporation of design and emotion into the curriculum for novice students may facilitate improved management of their work by establishing a correlation between emotions and decision-making in the design process.

1.2.5 DEFICIENCIES IN THE RESEARCH EVIDENCE

The impact of intrinsic emotional changes and the cognitive emotional system on the design process has been subject to limited research (Denton et al., 2004; Govers, Hekkert, and Schoormans, 2004). In addition, a limited number of design instructors acknowledge the ramifications and possibilities for emotional factors within the design methodology. This insufficient consideration towards emotions in the design process may stem from the notion that emotions are subjective and challenging to identify comprehensively. Moreover, a limited amount of extant design research is grounded in the viewpoint of students, and scant investigations have endeavoured to comprehend the extent and manner in which students' emotional fluctuations can impact the design procedure.

Although the research evidence is limited, prior studies that investigated the correlation between teaching and emotion have revealed that the outcomes of students' emotional responses are indicative of the teaching process's framework (Skinner, 1968). The study revealed that emotions have a positive impact on the process of exploration and creativity in design, as they facilitate the transfer and generalisation of knowledge across different disciplines. Furthermore, it was discovered that the presence of antagonistic emotions can motivate learners to surmount barriers in pursuit of their objectives and to consider alternative approaches to problem-solving within certain design methodologies (Scherer and Tran, 2001). The learning process of design studies had an impact on the intrinsic emotional changes and cognitive emotional systems of the students. Several emotions, including satisfaction, contentment, joy, and pride, were discovered to serve as motivators for students by bolstering their self-assurance in their abilities. This, in turn, facilitated their willingness to take risks, persevere through setbacks, and exercise sound judgement (Scherer and Tran, 2001). Consequently, an enhanced comprehension of these associations possesses the capability to assist pupils to enhance their design results by means of employing 'regulated emotions'. Overall, the correlation between emotion and design presents a promising avenue for research that can expand the existing body of knowledge and enrich the current literature on design studies.

1.3   JUSTIFICATION OF THE RESEARCH PROBLEM

As previously stated, it is widely acknowledged that contemporary design industry study into the design process is primarily concerned with controlling and facilitating the design process to produce a smoother and more successful end (Murty and Purcell, 2007; Stevenson, 2019; Takashi et al., 2007). In order to better understand how to build controllable design processes, academics have compared and contrasted different techniques in various creative processes (Cross and Sivaloganathan, 2004; Noble and Bestley, 2005; van Aken, 2005; Visocky O'Grady and Visocky O'Grady, 2017) and effective design research processes and methodologies were also investigated and analysed (Park et al., 2007); some of them developed innovative design processes for specific purposes (Park et al., 2007), and some processes were investigated and analysed (Park et al., 2007). Certain custom-made procedures were referred to as 'extrinsic' inputs since they helped to produce a more controllable and productive outcome. The majority of these investigations were prompted by commercial needs.

The methods of design management in college as well as how students cope with the time schedule and deadline for their entire design project were the subject of certain research investigations (Mann, 2004). Researchers also examined related topics, such as creative thinking, to develop more inventive design approaches, processes, and solutions (Fung, Lo, and Rao, 2005; Lawson, 2006; Sternberg, 1999). However, there are currently just a few available means to guide students through the design process in the most effective manner. Despite the fact that the approaches stated above have been given to them, a significant percentage of inexperienced design students/ designers at the entry level are unable to manage the entire design process and achieve success in their projects. Clearly, this demonstrates that there may be other factors that have a direct impact on the design process yet are frequently overlooked by academics.

1.4 DEFICIENCIES IN EVIDENCE

In studies of design, limited attention has been paid to the emotional changes that occur 'internally' as well as the cognitive emotional system that operates during the design process. There have only been a few studies that have offered support for this area of study (Denton et al., 2004; Govers, Hekkert, and Schoormans, 2004). The number of relevant research papers is quite small, and this element appears to have received little attention, particularly in terms of emotional shifts and the design process. Research is rarely conducted from the students' point of view to better understand how and to what extent their emotional shifts might influence the design process. Students can enhance their design outputs with 'controlled emotion' if they grasp the link between the two and work with instructors to learn more on that topic. The potential for further research is excellent, and this study will also make an important and valuable contribution to the existing pool of information to enhance the current standard of design education. As a result, a number of tailor-made methods have been developed that use 'extrinsic' inputs to facilitate more manageable and productive results. In all, it is generally recognised that the current research related to the design process mainly focuses on managing and facilitating it to achieve a smoother process and better outcomes (McClenaghan, 2007; Murty and Purcell, 2007; Takashi et al., 2007; Ting, 2007).

Some studies have investigated the design management methods used in design schools and explored how students deal with the time schedules for their design projects (Mann, 2004). Related studies have analysed creative thinking to develop more innovative design methods, processes, and solutions (Fung, Lo, and Rao, 2005; Lawson, 2006; Sternberg, 1999). Overall, a range of different management skills, including time management and risk management, have been used to prescribe and optimise the design process and the quality of the design outcomes. Designers have to learn the skills of allocating and choosing materials, as these are decisive factors in the design process. In addition, the costs, aesthetics, and quality of the materials have to be taken into consideration (Almendra and Christiaans, 2009; Scaletsky and Marques, 2009).

The changing nature of the design discipline forces designers to develop and evolve new approaches and strategies. Accordingly, rather than engage in conventional practices, designers strive to adopt more interactive models and to incorporate new ideas and creativity to achieve breakthroughs and innovation. Most design processes involve different sub-processes and components that interact to build a comprehensive system. Students are expected to be able to handle the design process using the extrinsic knowledge provided by the tutor. Yet many students find the design process difficult to master. Moreover, designers do not only receive information from objects, they also learn techniques, experience, and skills from the people around them. Because of the level of human interaction involved in design activities, for example, between the designer and his or her design teammates, the pattern of communication can affect the design process (Aken, 2005). The effects of human interaction on the decision-making of designers suggest that the design process involves not only rational and logical thinking but also the designers' responses to the external environment and the other people around them. As these responses to the external environment are manifest in emotional and affective states (Scherer, 1984), these 'intrinsic factors', which signify criteria that students encounter mentally, can be seen to influence the decision-making of designers.

1.5   THE AUDIENCE

This research study focuses on design studies for novice design students. Few studies have investigated the relationship between emotion and design from the point of view of novice students. Instead, most studies focus on developing ways of optimising design outcomes, such as gathering information on users' needs and effectively manipulating the design process. However, there is a lack of empirical research on the influence of novice design students' emotions.

Learning and achievement are critical for design students' career prospects. Pekrun (2006) suggests that students' achievement-related appraisals are important determinants of their emotions. This implies that academic activities in design often arouse intense emotions. Relevant achievement emotions include positive emotions such as enjoyment of learning, hope, and pride, while negative emotions such as anger, anxiety, and shame are caused by difficulties or failures in learning. All these emotions are functionally important as they influence students' academic motivation, behaviour, and performance.

While few empirical studies have focused on design students' emotions, existing research has investigated how emotions impair performance in complex or difficult tasks that demand cognitive resources (e.g. difficult mathematical tasks). An early study examined students' test anxiety as a kind of emotion (Zeidner, 1998, 2007) and produced cumulative findings that can inform design educational practice. The results showed that anxiety involves task-irrelevant thinking, which reduces task-related attention. In other words, students who worry about possible failure cannot focus their attention on learning. In comparison, positive correlations with academic achievement have been reported for positive emotions such as enjoyment of learning (Pekrun, 2006). However, Turner and Schallert (2001) found that negative emotions can have a positive effect on student achievement. For example, students' shame about failing an exam fuelled motivation to invest more effort in the future. Furthermore, feelings of boredom and hopelessness were found to have negative effects on cognitive resources, motivation, information processing and general academic performance (Pekrun, 2006). Overall, these findings indicate that emotions will have similar effects on the learning process of novice design students.

In general, novice design students are receptive to new ideas and have not been shaped by the experience and knowledge gained from their practice and tutors. Nonetheless, they gain a certain basic knowledge of design principles and the design process. They had also preliminarily developed their own design thinking, but did not yet possess much experience in handling design projects independently. They are capable of adopting an intuitive approach to design without recalling knowledge to control or manipulate the project for an expected outcome (Golec de Zavala et al., 2023; Keedy, 1998). However, as novice students do not have much experience of the design process, they are unable to make successive decisions based on their emotions (positive or negative) (Brissaud, Garro, and Poveda, 2003). Hence, novice design students were chosen as the investigating population in this study, which seeks to understand how emotional changes affect their design processes.

Novice students from bachelor degree programmes and all sub-degree design programmes (including those pursuing diploma studies, higher diplomas, and associate degrees) are the key target audience who would benefit from this study. The results of this study will enable those who are easily confused by the design process to better understand and manage their emotions. The results should also be of value to most novice students as they possess a little experience in handling a design process and have yet to develop their own independent thinking and practice.

1.5.1 DESIGN STUDIES FOR INEXPERIENCED DESIGN STUDENTS/DESIGNERS AT THE ENTRY LEVEL

The goal of this study is to investigate design studies for inexperienced design students/designers at the entry level in Hong Kong. Inexperienced design students/ designers at the entry level may not be completely prepared to build their own critical thought in terms of the design process, since the traditional secondary educational system usually stresses the outcomes rather than the process (Siu, 2003). The reason for selecting this particular student group is that they are uncontaminated by the experiences and knowledge obtained from their practices and teachers at the time of application. According to Brissaud, Garro, and Poveda (2003), the ability to take a more intuitive approach without recalling any prior information was demonstrated by certain inexperienced design students/designers at the entry level to manage or manipulate the project to get the desired design output. Some of them, on the other hand, had difficulty with the design process because they were unable to engage in consecutive discussions that were validated by decisions or bad feelings. As a result, by examining junior design groups, it was possible to gain a good understanding of how emotional shifts in designers may impact the design process overall.

The primary target audiences who would benefit from this research study include inexperienced design students from all sub-degree programmes (such as certificate studies, higher diplomas, and associate degrees) as well as students from the bachelor's degree programme. Inexperienced design students who are often confused by the design process will be able to grasp it and learn to regulate their emotions as a result of this experience. In addition, the results of this research study should be applicable to designers at the entry level of their careers, as the vast majority of them will have little experience in managing a design process and will not yet have developed their own independent thinking from the experiences and knowledge gained through their respective practices. Thus, in this research study, inexperienced design students/designers at the entry level of their careers should be considered to be the same target audience.

1.6   SIGNIFICANCE OF THE STUDY

The primary objective of this study is to investigate 'intrinsic variables' (Lum, 1997), which are emotional changes that occur in reaction to the design process that inexperienced design students/designers at the entry level may experience. As a result of the research, the emphasis of the study has been shifted away from the 'extrinsic' knowledge that assists professional designers in managing the development process and, instead, takes a creative approach to design. It investigates the link between emotions and the design process, as well as how students' emotions alter in response to their decision-making and how these factors have an impact on the complex design process. This project, which will employ both qualitative and quantitative techniques, has the potential to uncover additional critical elements that directly influence the design process and determine the effectiveness of the design output in the learning environment for inexperienced design students/designers at the entry level (Hall, Strangman, and Meyer, 2009).

This research study may have implications for the next generation of designers, design instructors, and perhaps the design industry itself. Inexperienced design students/designers at the entry level might, as a result, become aware of their emotional shifts as they progress through the design process. The reasons why inexperienced design students have difficulties with the design process and how emotions impact their decision-making and design processes may also be better understood by educators who work in the design field. They can look at various techniques that incorporate the concepts of 'design' and 'emotion' into regular design classes to assist the learning process of these students in general. This research investigates inexperienced design students' general understanding of the relationship between emotions and design processes and raises their awareness of the importance of the role of emotions in design; the topic of emotion and design is likely to be one of the themes covered in future design studies. At the beginning of their careers, inexperienced design students/designers at the entry level will, it is hoped, be better able to comprehend why emotional adjustments may improve the overall quality and standard of the design industry.

2 Theoretical Review

2.1 INTRODUCTION

'Design and emotion' has been a subject of study for more than a decade, with scholars devoting their time to various elements of the subject. To investigate the link between design and emotion and the emotions that it elicits, as well as to explain how emotion may be utilised effectively in design, a variety of research, models, and theories have been developed and implemented. Several words linked to emotional design have been mentioned in these studies, including emotion design and emotionalise design; in essence, synonyms of one another. The link between these concepts, as well as what they imply in their respective roles, and how these terms interact with one another within the larger picture of design and emotion, have been the subject of very few research studies. The purpose of this chapter is to present the ideas of emotional design, emotion design, and emotionalise design, as well as to examine their linkages from a fresh perspective. In light of these observations and analyses, the chapter proceeds to present a new conceptual model for identifying their distinctions by defining their meanings, in which both designers and users play a key role.

2.2 BACKGROUND OF EMOTION

2.2.1 THE PROCESS OF DESIGN DEFINED

It is possible to describe and investigate emotions in a variety of ways, and a variety of concepts have been proposed under the umbrella phrase 'emotion and design'. Discovering the links between different words linked to emotions and design, as well as their differences, begins with a thorough understanding of how emotion has been defined and developed over time. Plato (ca. 390 BCE/1955) is often considered to be the first thinker to use the term 'feeling' in the context of human history. According to this definition, emotion is defined as 'the immense pull of connectedness in love' as well as 'the intuition of significant ties between human beings'. Later, Aristotle, a prominent student of Plato's, claimed that emotions were a different type of judgement from other types. Several concepts in contemporary cognitive psychology, such as the appraisal theory, were developed as a result of Aristotle's theories, which will be addressed in more detail in the following paragraphs. With the help of Aristotle,

Charles Darwin, the famous English scientist of the nineteenth century, was the first to do a contemporary study on emotions. Darwin (2007) investigated the notion that emotion is a condition that is frequently linked with an adaptive goal. As patterns that formed to serve a purpose via learning, Darwin viewed emotional displays as having the potential to affect other social behaviours, such as human communication. As a result, they may enhance the chances of a species' survival. His knowledge of how external stimuli (e.g. events or objects) create emotional conditions served as the basis for the development of evolutionary theory.

William James (1884) believed that the body was necessary for experiencing emotion, based on a bodily-feedback perspective on the subject of emotion. Because he disagreed with Darwin's evolutionary theory, he proposed the peripheral hypothesis that emotion is accompanied by physical change. For example, an individual may declare they are in a state of terror because their heart is racing. Beginning in the 1930s, several academics began to investigate the link between an individual's emotional experience and elements in their immediate environment. A pervasive aspect of an experience that serves to unite and shape the experience, according to Dewey's (1934) book, *Art as Experience*, is described as emotion. This emotional characteristic cannot be traced back to a specific emotion or item. Rather, it is a gestalt produced by a variety of emotions experienced in the present instant, as well as memories of the past and interactions with the environment. Arnold and Gasson (1954) established the appraisal theory, which was founded on the concepts of Aristotle and Thomas Aquinas and developed over two decades. Affective states are defined by Arnold and Gasson as those psychological states that connect the exterior world of occurrences with the interior world of wants. Therefore, emotions are evaluations of an occurrence in relation to a certain purpose or objective (i.e. adjustments). With the concept of peripheral emotion, Schachter and Singer (1962) argued that physiological arousal may be linked to specific evaluations of a person's feelings. They created the two-factor hypothesis, which states that, for an emotion to be aroused, it is necessary for both peripheral physiological arousal and a cognitive assessment that identifies the arousal and links it to one emotion or another to exist simultaneously. Exactly one year after Schachter and Singer offered their theory, Tomkin (1995) released the first two volumes of his *Affect Imagery Consciousness* in which he brought together peripheral concepts and the facial expression studies outlined by Darwin to form a cohesive whole. Tomkin claimed in his affect theory that emotion was an affect programme that included both feedback from the body and conscious sensation and that it was responsible for our drives and motivations in life (Tomkin, 1995).

A greater number of surveys and debates were conducted throughout the two decades in an attempt to determine whether certain emotions are accompanied by distinct peripheral physiological response patterns. In comparison to a disgusted individual, do joyful individuals experience a higher increase in heart rate and skin temperature? It was not until Ekman (1972) and Scherer (1984) examined cross-cultural universalities of emotional manifestation that significant evidence for a fundamental-emotion theory was identified. Ekman and his colleagues proposed that particular facial muscle patterns are employed in the display of specific emotions; this was based on Tomkin's peripheral theory of facial expression. Furthermore, because people often exhibit basic emotions in the same way throughout cultures, facial expressions of fundamental emotions may be universal in nature. As a result, he believed that emotion offered a unique adaptive advantage in the transmission of critical information. From a cognitive standpoint, Scherer developed the fundamental-emotion theory, which asserted that emotion was the interplay of mood and behaviour and that being prepared for the behavioural reactions that are necessary to respond to both external and internal stimuli was essential. In his conclusion, Scherer (1984, p. 294) states that emotion is best understood as a series of interconnected, synchronised changes in the states of most of the five organism subsystems: motor nerves (in the central nervous system), sensory nerves (in the central nervous system), neuroendocrine system, autonomic nervous system, somatic nervous system (Bechara and Damasio, 2005). Each of the five subsystems has its own set of features or components, and the process of a single emotion episode includes the coordinated alterations of each of these states. The five emotion components for those mentioned organism subsystems were identified by Scherer as follows: cognitive component (i.e. appraisal and evaluation of stimuli and situations), neurophysiological component (i.e. bodily symptoms), motivational component (i.e. action tendencies), motor-expression component (i.e. facial and vocal expression), and the subjective feeling component (i.e. emotional experience). Frijda (1986) developed Tomkin's peripheral theory and linked it to Arnold's evaluation theory, which was published two years after Frijda. Frijda argued that the objective that was appraised in an event would generate an emotional response from the audience. Emotions, according to Frijda, are inclinations towards certain forms of interaction with objects and the world in which humans live. The embodiment of emotion is the expression of these inclinations in their many forms. To give an example, Frijda pointed out that the concealing and crouching motions associated with fear help to decrease the likelihood of being noticed or hurting someone. The expressive motions that represent the emotion, as a result, serve as a means of engaging with the surrounding environment. Based on the research of Scherer and Frijda, other psychologists began to regard emotion as a significant factor in the understanding of human behaviour, and emotion was recognised as being crucial for understanding many core phenomena in some subdisciplines of psychology, such as social psychology.

During a conference conducted in 1990, which was the forerunner of the McDonnell Pew Program in Cognitive Neuroscience, the latest discoveries in the field of emotion research were discussed. After listening to the debate, Davidson and Cacioppo (1992) established emotion as a self-organising and integrative state that is coherent across many response systems, such as social-psychobiological-behavioural-informational factors. Meanwhile, Tassinary and Cacioppo (1992) investigated how design, being one of the ways in which information is processed, may affect the views of consumers (self-referencing) as they go through their information-receiving routines. Consequently, designers must recognise that design (through the use of both visual and linguistic methods) may elicit emotional responses (the perceptions of consumers). Previous studies offered a considerable body of evidence that emotion has an impact on cognition, even though few cognition models address emotion. During the course of his book *Experienced Cognition*, Carlson (1997) emphasises the distinction between emotional and affective states. In his definition, emotion is characterised by brief, acute waves of sensation that arise without the need for conscious effort or contemplation and are frequently accompanied by developing physiological changes, such as an increased pulse. When it comes to emotional state, on the other hand, this is described as a more prolonged and less strong emotional influence. Even though Carlson (1997) and Dewey (1934) employed different terminology, they both agreed on the existence of two forms of emotion, that is, one in which the emotion is brief and reflexive and another in which the emotion is prolonged and reflective. As a follow-up to the work of Tassinary and Cacioppo, Creusen (1998) emphasised that emotion may be a significant component in customers' purchase decisions, as it influences their decision-making when selecting goods and services. According to the results of these studies, emotion is not only a response to both external and internal stimuli, but it also serves a variety of functions, such as the evaluation of physical phenomena, support and supervision, preparedness and direction of operation, communication of response and actual behaviour, and supervision of the inner state of the organism. This infers that emotion has an impact on human behaviour, which includes information processing and decision-making.

2.2.2 DIFFERENTIATION OF EMOTIONS

Different techniques for distinguishing emotions were proposed in conjunction with the diverse definitions of emotion that were found in psychological research. Ekman (1972) defined seven different emotions based on facial expressions: surprise, joy, sadness, disgust, fear, rage, and contempt. Each emotion is linked with a certain type of facial expression; for example, rage is characterised by a fixed gaze, tightened brows, compressed lips, and forceful and rapid motions. Ekman and Friesen (1975) developed a formalised technique that was consequently used by many other academics, who compiled comparable collections of feelings that are now referred to as 'basic emotions'. Ekman's idea was also proposed by Plutchik (1980), who also created a theory based on the eight basic human emotions of joy, acceptance, fear, submission, sorrow, disgust, rage, and anticipation. He said that these eight fundamental emotions were the source of all other human emotions. The development of secondary emotions – those feelings that are not included in the eight main human emotions – would be derived from the mixing of the eight emotions to show the responses of a person to stimuli. Arnold (1960) suggested that each emotion was the result of a unique assessment, which he attributed to the effect of the appraisal viewpoint of emotion research. He and his colleagues not only concentrated on the methods for distinguishing different emotions but also attempted to explain how emotions were elicited. Although each stimulus can trigger a wide range of emotions, each evaluation type is associated with a unique feeling. To separate emotions, Ortony, Clore, and Collins (1988) merged the ideas of 'fundamental emotions' developed by Ekman with the concept of 'preceding evaluation' developed by Arnold. Following the application of the fundamental principles to identify the most evident emotions, Arnold built a platform on which to investigate the links between the evaluation and the most obvious emotions. For example, when a person is anticipating a desirable occurrence, their enjoyment may be plainly witnessed, and they consider the event as the assessment of the emotion 'joy'. At that time, other psychology researchers asserted that emotions were intertwined and that specific emotions (e.g. anger and annoyance) were more similar than others (e.g. happiness and sadness, as well as boredom vs inspiration). Hence, they recommended that emotions be described and differentiated according to their fundamental characteristics (Wundt, 1905) to better understand them. This 'underlying dimension' has been discovered in a variety of research efforts, including emotional words (Averill, 1975), facial expressions (Gladstones, 1962), and self-reported emotions (Averill, 1975; Mehrabian and Russell, 1977). Since the 1980s, the dimensions have been referred to as 'pleasantness' and 'activity', respectively. It is possible to be unpleasant (e.g. unhappy) while also being nice according to the dimension of 'pleasantness' (e.g. happy). Arousal is described as physiological arousal (Clore, 1994), and it can range from calm (e.g. content) to enthusiastic (e.g. ecstatic or euphoric). Russell (1980) identified eight categories of emotions (neutral excited, pleasant excited, pleasant average, pleasant calm, neutral calm, unpleasant calm, unpleasant average, and unpleasant excited) that may be categorised using the two dimensions.

Taking inspiration from the concept of differentiating emotions through the use of underlying dimensions, some psychological scholars re-examined the concept of 'basic emotions' and suggested a more effective approach that distinguishes emotions from one another, based on several components (Lazarus, Kanner, and Folkman, 1980), which included behavioural responses, expressive responses, physiological reactions, and subjective feelings. Behavioural responses are the activities or behaviours that one engages in when experiencing an emotion, such as running, seeking contact, or other similar actions or behaviours (Arnold, 1960). The term 'expressive responses' refers to the facial, verbal, and postural projections that occur in conjunction with an emotion. To portray each emotion, a certain sequence of facial expressions must be used. Formalised responses were defined by Ekman (1972). According to Lazarus (1993), 'Physiological responses' relate to a range of physiological manifestations that occur as a result of emotional experiences, such as a rise in heart rate and sweating. In conclusion, 'subjective feelings' relate to the conscious knowledge of one's own emotional state, also known as the 'subjective emotional experience', which is distinct from an objective emotional experience. Psychological researchers can identify emotions from different points of view, based on these components, and the expressions of emotion can be assessed as a result of these considerations. However, because the feelings of individuals were abstract and difficult to be differentiated from one another based on the fundamental emotions, the researchers were unable to reflect all of the emotions experienced by the participants in the studies.

As a result of the techniques described above, an emotion will be differentiated based on its expressions, prior evaluations, and underlying characteristics. However, among these, the difference in the underlying dimensions is the most efficient and thorough method of guiding users to describe their emotions. Because emotions are interconnected and some emotions are more similar (e.g. anger vs irritation) than others (e.g. boredom vs inspiration), they are best described and differentiated by users who are familiar with the underlying dimensions (e.g. anger vs irritation). The participants in the in-depth focus group of the second empirical study in this series of research investigations are anticipated to explain their emotional shifts as a result of a wide range of interconnected emotions, as well as how these emotions impact their decision-making and design processes. As a result, the differentiation of emotion based on fundamental aspects is used in this situation.

Each of the techniques for distinguishing between emotions that have been discussed has its own set of advantages and disadvantages. As a result, the implementation of the research should be determined in accordance with the goals of the research. Apart from the diverse psychological explications of emotion, scholarly research has proposed multiple methodologies for distinguishing between emotions. Ekman (1972) categorised seven discrete emotions based on their corresponding facial expressions: surprise, joy, sadness, disgust, fear, anger, and contempt. Distinct facial expressions are linked to specific emotions, such as anger which is characterised by a steady gaze, furrowed brows, pursed lips, and forceful and energetic gestures (Ekman and Friesen, 1975). Numerous scholars have adopted this methodology and assembled comparable collections of emotions that are commonly recognised as 'fundamental emotions'. Plutchik (1980) conducted a review of Ekman's work and formulated a theoretical framework centred on eight fundamental emotions that are inherent in human beings. These emotions include joy, acceptance, fear, submission, sadness, disgust, anger, and anticipation. According to Plutchik's theory, the entirety of human emotions can be traced back to eight fundamental states. As such, any emotions that are considered 'secondary' would be the result of a combination of two or more of these primary states and would ultimately reveal an individual's reaction to a given stimulus. Arnold (1960) postulated that individual emotions are the result of distinct appraisals, drawing from the appraisal theory of emotion. Rather than only focusing on the way to distinguish between emotions, Arnold also tried to explain how emotions are elicited. While it is true that a given stimulus may provoke a variety of emotional responses, it is also the case that each distinct type of appraisal is associated with a particular emotion. Ortony, Clore, and Collins (1988) integrated Ekman's notion of 'basic emotions' (i.e. basic emotions theory) and Arnold's concept of a 'preceding appraisal' (i.e. appraisal theory) in order to distinguish between emotions. The researchers utilised the fundamental emotions posited by the theory of the basic emotions to discern the most salient emotions and subsequently utilised this groundwork to investigate the interconnections between appraisal and emotion. The researchers noted that the appraisal of the emotion 'joy' can be readily observed through the pleasure experienced by an individual in response to a desirable event. During a concurrent period, a group of scholars in the field of psychology posited that emotions exhibit interconnectedness and that certain emotions share greater similarities with each other (such as anger and irritation) in comparison to other emotions (e.g. boredom versus inspiration). Ortony, Clore, and Collins (1988) proposed a description and differentiation of emotions based on the underlying dimensions of emotion as suggested by Wundt in 1905. Several psychologists have incorporated Wundt's ideas of 'underlying dimensions' into their research, resulting in the identification of various underlying dimensions across different studies. These dimensions include emotion words (Averill, 1975), facial expressions (Gladstones, 1962), and self-reported emotions (Mehrabian and Russell, 1977). Prior to the 1980s, the afore-mentioned dimensions were commonly categorised as pleasantness and activation.

The concept of pleasantness encompasses a spectrum of emotional states, ranging from negative affect such as unhappiness to positive affect such as happiness. On the other hand, activation pertains to a level of physiological arousal as defined by Clore (1994) and spans from a state of calmness, such as contentment, to a state of heightened excitement, such as euphoria. According to Russell's (1980) classification, emotions can be categorised into eight distinct types based on the following specific dimensions:

1. Neutral excited (e.g. aroused, astonished, stimulated, surprised, active, intense)
2. Pleasant excited (e.g. enthusiastic, elated, excited, euphoric, lively, peppy)
3. Pleasant average (e.g. happy, delighted, glad, cheerful, warm-hearted, pleased)
4. Pleasant calm (e.g. relaxed, content, at rest, calm, serene, at ease)
5. Neutral calm (e.g. quiet, tranquil, still, inactive, idle, passive)
6. Unpleasant calm (e.g. dull, tired, drowsy, sluggish, bored, droopy)
7. Unpleasant average (e.g. unhappy, miserable, sad, grouchy, gloomy, blue)
8. Unpleasant excited (e.g. distressed, annoyed, fearful, nervous, jittery, anxious)

Informed by the notion of emotion segregation based on fundamental dimensions, a group of scholars in the field of psychology conducted a review of the notion of 'basic emotions' and proposed a more comprehensive approach that employs multiple criteria to differentiate between emotions (Lazarus, Kanner, and Folkman, 1980). These criteria encompass behavioural reactions, expressive reactions, physiological reactions, and subjective feelings. Behavioural reactions refer to actions or behaviours that are associated with the experience of an emotion, such as the act of fleeing or seeking contact (Arnold, 1960). The phenomenon of expressive reactions pertains to the manifestation of emotions through distinct patterns of facial, vocal, and postural expressions. Each emotion is characterised by a unique set of expressions that accompany it (Ekman, 1972). Physiological reactions refer to a range of bodily responses that co-occur with emotional experiences, such as heightened heart rate and perspiration levels (Lazarus, 1993). Subjective feelings refer to the conscious perception of an individual's emotional state, which denotes their subjective emotional experience (Frijda, 1986). These constituents facilitate psychologists in discerning emotions from diverse viewpoints and quantifying the expressions of emotion. Nonetheless, this methodology of categorising emotions based on the fundamental emotions is limited in its ability to encompass the entirety of the human emotional experience.

The aforementioned methodologies demonstrate that emotions can be distinguished based on their expressions, such as antecedent evaluations and fundamental dimensions. The utilisation of differentials pertaining to the underlying dimensions is deemed the most efficacious and all-encompassing approach that may assist users to articulate their emotional states. Given the interrelatedness of emotions, it is optimal to describe and distinguish them based on their underlying dimensions, particularly as certain emotions, such as anger and irritation, exhibit greater similarity than others, such as boredom and inspiration. The participants in the focus groups of the second empirical study were requested to articulate their emotional transformations in relation to a range of interconnected emotions and their impact on the decision-making process and design procedures. The preceding research on the differentiation of emotions has served as a source of inspiration for design professionals in their efforts to classify emotions within the realm of design research. The categorisation of emotions based on fundamental dimensions has been extensively utilised in the field of design and emotion research. However, each of the aforementioned methodologies possesses its own set of merits and demerits. Therefore, it is imperative for research studies to employ various methods to distinguish emotions based on their research objectives.

2.2.3 EXTENSIVE DISCUSSIONS ABOUT EMOTION IN OTHER STUDIES

Since the 1970s, the study of emotion has extended into a variety of fields, including philosophy, sociology, and economic science as a result of the inspiration gained from psychological research. Emotion has mostly been associated with subjects like information processing and decision-making, according to academics. To better understand why individuals prefer to find meaning in items, Csikszentmihalyi and Rochberg-Halton (1981) investigated (from psychological and symbolic viewpoints) how emotion may assist humans in comprehending the environment. People's emotional and mental states, as well as their moods, were observed to have an impact on the techniques by which they processed information. Baudrillard (1981), from a philosophical standpoint, investigated how emotion influenced information processing, as well as how it affected customer behaviour. He concluded that changes in customers' emotions would have an impact on their unconscious wants. If a product is to remain competitive, it will be necessary to increase the attractiveness of the product (i.e. design outcomes) to satisfy the unconscious wants of customers. The link between emotional concerns and information processing was explored in detail by Picard (1997) in her book on media studies, *Affective Computing*. The research revealed that emotions are subjective experiences that have an impact on both the human capacity to comprehend information and the human reaction to an event while investigating ways to optimise computer-controlled systems. Picard's research led her to the conclusion that to increase information processing between users and computerised systems, the design of computerised systems should incorporate emotional skills. With the introduction of this notion, an increasing number of academics began to consider the applicability of these emotion theories. For example, research by economists uncovered that it is difficult to advertise items in marketplaces when there are few distinctions between them. Thackara (1997) concluded that, in situations where products are similar in terms of their technical characteristics, quality, and price in the current market, innovative design is required to increase the attractiveness and market competitiveness of the new design and to enrich the users' experiences. Pine and Gilmore (1999) researched emotion theories and argued that goods should include users' experiences to win market share. Jensen (1999) went on to say that human emotions will serve as an important connection between consumers and goods in the future. Instead of focusing on the function of things, consumers would shift their attention away from purchasing products and towards appreciating the sensations and feelings that products communicate. The importance of emotions in the development of new designs and communication with customers was further emphasised by Schmitt (1999). As a result of economic research on how emotional issues impact the economy and goods, some design professionals began to observe what customers wanted in the future and examined how emotional concerns were included in the design process.

2.2.4 The Study of the Relationship between Emotion and Design

Some researchers put their emphasis on emotion, which prompted design experts to investigate the link between design and emotion. The scholarly pursuit of emotions has motivated professionals in the field of design to investigate the correlation between design and emotion. During the 1950s, the design industry placed emphasis on utilitarian principles and aesthetics that were primarily focused on functionality. Functionalism is a theoretical perspective that posits that the constituent components of a system can be comprehended solely by examining the roles they fulfil in relation to the entirety of the system (Walker, 1989). The concept of functionalism was utilised to investigate the efficiency and efficacy of the design process. Scholars in the field of design have endeavoured to establish efficacious and clearly defined methodologies for designers to employ in the development of their designs. In response to the evolving nature of the design discipline during the 1950s, designers were compelled to devise novel media and tactics. Following the 1960s, design methodologies underwent a transformation, departing from traditional prototyping practices and embracing interactive models that integrated novel concepts to foster advancements and originality. It is currently posited by scholars that the process of design can be segmented into several distinct stages. The analysis–synthesis model, introduced by Archer in 1965, outlines a design process that encompasses three primary stages: analytical, creative, and executive. According to Archer's proposal, the process of problem analysis, which occurs during the analytical stage, is separate from the process of design synthesis, which takes place during the creative and executive stages. Archer further posited that designers must engage in a comprehensive analysis of the problem and establish a complete definition of it before proceeding to synthesise a solution.

Jones (1984) expanded upon the analysis–synthesis model by incorporating Archer's logical methodology and integrating it with intuition, experience, and a rigorous logical treatment. The systematic design concept that he developed posited that designers ought to employ both rational analysis and imaginative thinking in the course of addressing problems. The author posited an alternative design process comprising three distinct stages – analysis, synthesis, and evaluation – with the aim of offering a framework for the implementation of the aforementioned concept. The proposed guideline suggested that designers should incorporate practical constraints and rational reasoning while generating concepts and resolving problems. The preliminary design prototypes indicated that the conception of the design procedure was initially centred on logical progression and sequential methodologies. The design processes were systematically executed with a solid foundation of logical justifications. Luckman (1967) expanded Jones' systematic design concept by incorporating insights from designers' actual work processes during the late 1960s. Luckman's study proposed that the practical application of the design process is characterised by a non-linear progression and that the various stages of the design process ought to be iterated at varying degrees of granularity. The iterative nature of the systematic design concept, comprising analysis, synthesis, and evaluation, was underscored by Luckman in the context of practising design. To attain potential solutions, it is necessary to carry out numerous iterations of information translation processes, such as requirements, constraints, and experience evaluations, throughout the entire design process, from this standpoint. The topic of design focuses on design outputs, designers, and the design process (Goldschmidt, 1999). In the 1950s, design was characterised by an emphasis on functionality, utilitarianism, and aesthetics. According to the notion of functionalism, a system's components can only be characterised in terms of the roles they play in relation to the whole (Walker, 1989).

The majority of prior research on the design process has been approached from the viewpoint of the designer. During the 1980s, scholars in the field of design commenced an examination of consumer trends. They suggested that the principle of 'form follows function' in modernism was no longer sufficient to meet the needs of consumers/users, given the increasing research attention being paid to emotions in other disciplines (Walker, 1989). Currently, design plays a crucial role in markets that are dominated by consumers in the United States and other nations. Consumers residing in these markets have observed that conventional functional design tends to be uninteresting. However, they have also noted that incorporating innovations that cater to their psychological needs can enhance the functionality of consumer durables. By taking into account the psychological requirements of consumers, designers have the ability to modify the design process and resulting designs to align with the genuine needs of users. The efficiency and efficacy of the design process were investigated under the functionalist paradigm. Design researchers attempted to identify successful strategies for designers to use as a well-defined approach for their creations. Designers were compelled to invent and adopt new mediums and techniques in response to the evolving structure of the design discipline. Since the 1960s, design processes have shifted away from the traditional prototype in favour of a more divergent and interactive approach that incorporates new ideas and creativity to generate breakthroughs and innovations. The majority of scholars feel that design processes can be classified into several stages. Archer (1965) was the first to propose the analysis–synthesis model for design processes. The approach divided design into three distinct stages: analytical, creative, and executive. Archer differentiated between issue analysis (during the analytical stage) and design synthesis (in the creative and executive stages). According to his methodology, designers must thoroughly analyse and describe the problem and then synthesise a solution. Archer initiated early research into the design process, concentrating on cognitive stages and rational methods. By emulating Archer's rational techniques, Jones (1984) defined the analysis–synthesis model and established a connection between it and intuition, experience, and rigorous logical treatment. As a result, he created the notion of 'systematic design', which advocated for designers to combine logical analysis with creative thinking during the problem-solving process. To offer designers guidance on how to apply the notion, he developed another design method composed of three stages: analysis, synthesis, and assessment. Designers would generate concepts and solutions while taking into account real-world constraints and using logical judgement. According to earlier design models, the initial design process was largely rational with linear procedures. These step-by-step design procedures were based on sound rationales. However, in the late 1960s, Luckman (1967) argued that the design process is a non-linear one in practice and that the phases in the design process should be repeated at various degrees of design complexity, based on Jones' idea and observation of designers' actual working processes. Luckman emphasised that, in design practice, the three phases of systematic design (analysis, synthesis, and assessment) would be repeated in cycles. Numerous information translation procedures (including requirements, limitations, and experience) were repeated throughout the design process to arrive at possible solutions. Consumer societies began in America and expanded across the rest of the Western world.

With the progression of research, diverse design methodologies were proposed to align with the pragmatic design scenario prevalent in the 1980s. The practice of professional design encompasses a diverse range of fields, including but not limited to industrial design, graphic design, and interior design. Hence, it is anticipated that designers will address the diverse design demands of their vocation through innovative, comprehensive, and all-encompassing approaches. The design process has become a prominent subject in design research, as it constitutes a fundamental aspect of designers' daily practise. According to Hillier, Musgrove, and O' Sullivan (1984), design primarily involves problem-solving. The proposition posits that the process of problem analysis during design encompasses not only cognitive activities but also incorporates the designers' personal knowledge and imagination. As per their account, design can be characterised as an activity involving conjecture and analysis. During the conjecture phase, designers engage in cognitive processes to conceptualise an outline and subsequently employ artistic techniques, such as analogy, metaphor, and sudden insights, to generate novel ideas. During the analysis phase, designers employ a logical and scientific approach to examine the implications of a novel concept in light of the diverse demands posed by the design challenge.

As additional research was conducted, a variety of design techniques were proposed to address the practical design issues seen in the 1980s. Currently, designers engage in several professional activities, including industrial design, graphic design, and interior design. As a result, designers are expected to be creative individuals who create objects that integrate and holistically address different design criteria. The design process is the daily workflow that designers employ and has evolved into a significant area of design study. Hillier, Musgrove, and O' Sullivan (1984) stated that the primary purpose of design activities is to solve problems. Not only was problem analysis a cognitive exercise, but it also incorporated the designer's personal expertise and creativity. As a result, they asserted that design is a 'speculation/analysis activity'. In the speculative mode, designers think cognitively about the concept outline and then generate additional ideas using creative processes, such as analogy, metaphor, and unexpected flashes of insight. In the analysis mode, designers employ logical scientific reasoning to determine the impact of a novel concept on the design problem's numerous needs. Akin (1984), based on the conjecture-analysis model, emphasised that design was a process of issue exploration. Designers do not limit themselves to physical and quantitative studies while structuring their quest. Additionally, individuals make judgements and evaluations, based on the personal information and abilities they have gained from their own experiences. As a result, Akin argued that the design process is divided into three areas of reasoning: the objective domain (i.e. the problem's actuality), the representational domain (i.e. how reality is seen), and the construction domain (i.e. the reality of the solution).

Akin (1984) emphasised that the process of design involves problem exploration, according to the conjecture-analysis model. Designers employ a variety of methods to structure their research, beyond the use of physical and mathematical analyses. In addition to these methods, designers draw upon their personal knowledge and skills, developed through their own experiences, to make informed judgements and evaluations. Akin postulated that the process of designing encompasses three distinct domains of reasoning, namely the objective domain that pertains to the actual problem at hand, the representational domain that concerns the perception of reality, and the construction domain that deals with the actual creation of the design (i.e. the reality of the solution).

Csikszentmihalyi (1996) proposed an alternative model of the design process that incorporates multidirectional research and thinking processes, drawing from design practices. Csikszentmihalyi's design methodology encompasses five distinct stages: preparation, incubation, insight, evaluation, and elaboration. The first stage involves identifying a set of intriguing problems that stimulate curiosity. During the incubation stage, ideas are subconsciously processed, leading to the formation of unique connections. The insight stage marks the point at which the puzzle pieces begin to fall into place. In the evaluation stage, the most valuable insights are selected and deemed worthy of further exploration. Finally, the elaboration stage involves fleshing out the chosen insights into a fully realised design (i.e. turning the insight into something real). The proposer posited that the aforementioned design process model serves as a beneficial point of reference for designers due to its methodical organisation. Nevertheless, certain academics have posited that the standardisation of the design process may not be advisable. As per the findings of Austin and Devlin (2003), the process of designing is essentially a journey of exploration and revelation. Although the objective of the procedure may be predetermined, the means of attaining it remains uncertain. According to Austin and Devlin, the process of creative problem-solving lacks a sequential and linear structure and is characterised by the absence of clearly defined steps due to the frequent involvement of new inputs and innovative thinking. Best (2006) underscored the impact of varying client and user needs on the execution of the design process. Best posited that designers can refine their designs through trial-and-error methods. However, several experts claimed that the standardisation of design procedures was impossible. According to Austin and Devlin (2003), design is a discovery process. The objective of the process may be known, but the means by which it will be accomplished is uncertain. They argued that creative issue resolution should be an ad hoc, non-linear process with no clearly defined phases, as it typically incorporates novel sources and inventive ideas. Best (2006) emphasised the importance of the distinct demands of clients and users on the functioning of the design process. Trial-and-error approaches assist designers in modifying their design processes and methodologies. As a result, the design process has become increasingly unlike those described in previous decades. Certain design approaches offered were multidirectional, non-linear, and centred on the demands of users/consumers. Changes in technology, society, culture, and economy, as well as other variables, all affect the function of designers in the design process (Ho and Siu, 2009b). As a result, beginning in the 1990s, designers were compelled to experiment with new paths that connected design and emotion.

Recent conceptualisations of the design process have exhibited a tendency to deviate from prior models. Several design processes have been proposed that involve multiple directions and non-linear processes with a particular emphasis on addressing the requirements of users or consumers. In contemporary times, the role of designers in the design process is observed to be influenced by various factors, such as technological advancements, social dynamics, cultural shifts, and economic developments (Ho and Siu, 2009b). Since the 1990s, designers have endeavoured to explore novel avenues, such as design and emotion, in order to enhance the efficacy of the design process and the calibre of its results.

Several academic researchers have conducted extensive investigations into the evolution and modifications of design, the design process, and the requirements of users and consumers. Their findings indicate that a range of technological, social, cultural, and economic factors can significantly impact the role of designers in the design process (Hummels, 1999). Among these factors, the conspicuous influence of technology on the design process stands out. The advancement of technology has led to an escalation in intricacy and challenges associated with producing exceptional design results. Due to the inadequacy of current designs in meeting the demands of users and consumers, there is an increasing necessity to integrate emotions and technology in design to enhance its capacity to satisfy human emotional requirements (Rosella, 2002). The current research trend emphasises the interaction between design outcomes and users, revealing that the advancement of technology has led society to prioritise the provision of human experience services over functional products (Sanders, 1999). Moreover, the cultural milieu has undergone a transformation in response to technological and social shifts, with a shift in emphasis from communal to individualistic concerns. Consequently, there exists a significant need for designs pertaining to identity, which typically incorporate intense personal sentiments from either designers or clients, within the realm of communication design. The designs elicit a sense of identity through the utilisation of symbols and forms that evoke a potent emotional response in the user (Ben-Peshat, 2004). Simultaneously, design outcomes that are informed by emotions offer product differentiation and aid companies in acquiring fresh market opportunities amidst the evolving economic landscape. Design has been incorporated into the latest business strategies of organisations that aim to enhance their competitive edge by creating more cohesive products.

Walker (1989) asserted that, in light of the growing focus on emotion in other areas, as noted above, modernism's dictum 'Form follows function' could not satisfy consumers/users. The design was critical in the USA and other countries' consumer-dominated markets. Customers in these areas expressed dissatisfaction with functional designs but noted that innovations that incorporated consumers' psychological requirements could improve the utility of consumer durables. Simultaneously, academics discovered that some elements, such as technological, social, cultural, and economic considerations, would impact both the function of designers in the design process and their emotions. Among all of these elements, the most evident was the impact of technical factors on the design process. The complexity and difficulty of producing great design outputs have risen as a result of technological advancement. Hummels (1999) emphasised the inability of present designs to meet the demands of users/consumers. Rosella (2002) concurred with Hummels and proposed integrating emotions and technology into all design processes to enhance design practices for meeting human emotional requirements. As a result, the new product trend is centred on the way design outputs and users interact. According to Sanders (1999), the progress of advanced technology has forced society's productivity focus to shift away from functional items and towards human experience services. On the other hand, cultural factors reflect technical and social developments. The emphasis of the cultural settings has shifted away from the community and towards the individual. Because identity designs are frequently characterised by strong personal feelings (either from designers or buyers), they were in high demand across many cultural groups in society. According to Ben-Peshat (2004), these designs foster a feeling of identity through the use of symbiotic meanings and serve as indicators of society's powerful emotional influence. Simultaneously, emotional design results can differentiate products and enable businesses to capitalise on new market possibilities in a changing economic climate. To help businesses achieve their objectives, design has become an integral component of their new business strategies, particularly for those seeking a competitive edge through more integrated products.

Cooper (1999) pioneered research on the link between design and emotion. He contended that the bulk of technical gadgets (e.g. videocassette recorders, vehicle alarms, and software programmes) made users feel inadequate and irritated as a result of poorly designed user interfaces. Following Cooper, several design academics examined the emotional aspects of design. Overbeeke and Hekkert (1999) coined the term 'design and emotion' for the first time. Their research on design and emotion aimed to produce tools and approaches that would assist a designer in developing an emotionally worthwhile product–user connection. As additional research was conducted, a network for discussing design and emotion among design researchers became necessary. As a result, the Society for Design and Emotion was founded in 1999 (Demset and Hekkert, 2009). It established a worldwide community of scholars, designers, and businesses interested in and experienced with design and emotion.

Since 1999, further research has emerged on the subject of design and emotion. Several academics expanded on this area of study by examining it through the lens of several disciplines and themes, including product experience, user experience, and the three levels of design. Overbeeke and Hekkert (1999) argued for the critical nature of product experience. The emphasis on emotion is considered to add to the ease and efficiency with which designs are used. Enders (2004) proposed that including the idea of user experience would aid in optimising design results. Suri (2003) noted that many goods (or design outputs) on the market today have characteristics that include technological functionality, pricing, and quality. To enhance design results, designers need to create more exceptional outcomes that meet customer demands. Designers should gain an understanding of their consumers' experiences, investigate design concepts that meet their needs, and include emotional considerations in the design process. Meanwhile, certain external variables also affect the designers' emotions and decision-making. It would be beneficial to conduct an extensive variety of studies on the link between the external environment and emotions to detect these variables in the future. It is recognised that certain elements, such as social, cultural, and technical evolution, cannot be controlled by designers and have an effect on their emotions and decision-making during the design process. Additionally, by integrating consumers into the design process, emotion and experience may be promoted. Norman (2004) expanded on the 'three levels of design' notion, where he classified designs into three categories, based on the user's/consumer's responses to the design's consumption: visceral, behavioural, and introspective. The visceral level relates to the user's/consumer's initial reaction to the design output, whereas the emotional response is intuitive. The behavioural level refers to the behaviours taken by users/consumers in response to the emotions evoked by design results. The reflective level refers to the reflections of users/consumers on their consumption experiences. Norman's work elucidates how designs affect users/consumers emotionally.

In the years that followed, several design researchers emphasised an experimental approach to established design and emotional ideas. To further examine the link between emotion and many parts of the design, including the design process and design objects, Enders (2004) used an experimental design process model dubbed LEONARDO to investigate the application of emotion theories during the design process. Several researchers examined how users' prior experiences affected their current views and feelings. According to Demirbilek and Sener (2002), story character characteristics should be incorporated into developed goods to elicit favourable emotional responses from consumers. As with Demirbilek and Sener, Spillers (2004) sought to investigate how the usability of a product affected the user's experience, attitudes, expectations, and motives (emotions). He found that emotion was significant in that it influenced how users/consumers understood, investigated, and assessed the consequences of a design.

Numerous classifications of ideas and research have been proposed, based on various perspectives of design and emotion studies. Desmet and Hekkert (2009) proposed one of the most comprehensive and contributive/indicative methods in an editorial for the special edition of the *International Journal of Design*, titled 'Design and Emotion', which commemorated the International Design and Emotion Society's 10th anniversary. Desmet and Hekkert (2009) began developing techniques for systematising the many views on design and emotion research. The authors reviewed prior studies and concluded that the user–emotion link extended beyond the product to concerns about retail price, service, and brand. They discovered that certain research techniques integrated the user–emotion link and associated tools to assist designers in designing to invoke emotion. Desmet and Hekkert (2009) distinguished themselves from prior design and emotion research in two ways and presented a cross-sectional assessment of the field's advances as follows:

1. The research's stated goals are to 'improve design quality' (i.e. to make products more engaging, more authentic, and easier to use).
2. The techniques adopted by the researchers in their investigations (i.e. user-based, designer-based, research-based, and theory-based).

They discovered that the studies conducted under the rubric of 'making things more engaging' were primarily concerned with the user's experience. On the one hand, design academics are always looking for new and creative methods to make their designs more fascinating and engaging. Some design researchers, on the other hand, believe that emotions contribute to the ease and efficiency with which a product is used and, accordingly, concentrate their efforts on the issue of 'increasing ease of use'. Other designers have voiced their view that most commercial designs are based on the simple enjoyment of 'making goods more authentic', principles that they believe are supported by design academics. Additionally, the methodologies of 'designing for emotion' were identified and implemented into the product development process, as well as the roles of the various actors, which included engineers, designers, and marketers. The techniques taken by Desmet and Hekkert were consequently distinct from those taken by existing research. Studies that employed a 'user-based' approach focused on the feelings and experiences of users, and their aims and aspirations served as the inspiration for the study. Designers were observed to be authors under the 'designer-based' approach, and their creations were viewed as a means of expressing ideas between parties. For research-based techniques to disclose the link between design decisions and emotional reactions, it was necessary to assess emotional responses. The research that provided insights into users'/consumers' behaviour to optimise designs examined 'theory-based' methods. However, few design and emotion studies went on to explore the real relationships that exist between designers, users/consumers, and the design outputs and how the function of emotion impacts the design process, consumption, and the interaction between designers and users/consumers (or vice versa).

Ho and Siu (2009a) proposed categorising different research efforts and theories into three main types by relying on the positions of designers, design outcomes, and users/consumers in the design process, as well as consumption (i.e. 'users/consumers-driven', 'designer-driven', and the relationship between users/consumers, designers, design outcomes, and other factors). Among other purposes, this strategy is used to improve comprehension of the key role played by emotions in the interactions between the roles of designers, design outcomes, and users/consumers. An in-depth analysis of the related literature will now be presented.

2.3 DIFFERENT TYPES OF CONCENTRATION IN DESIGN AND EMOTION RESEARCH

2.3.1 USERS/CONSUMERS-DRIVEN RESEARCH

The research on users/consumers-driven studies (i.e. the relationships between users/consumers and design outcomes) was the first type of study explored by scholars in the field of design and emotion, of the three main types of studies based on the roles of designers, design outcomes, and users/consumers in the design process and consumption. They felt that if designers had more understanding of the user/consumer experience, it would be simpler for them to create better products and services (Desmet and Hekkert, 2009).

Design researchers have drawn on emotion theories to generate notions that will enhance the outcomes of their designs (i.e. products). When Gaver (1999) developed the concept of 'culture probes', they were referring to a research approach that allowed designers to gain contextualised and deep insights from the experiences of users/consumers. Emotion products (i.e. designs that incorporate emotional concerns) were believed to stimulate an 'interaction style' that was more intuitive and sensitive, and it was beneficial for the designer to interact with users/consumers to gain their insights as feedback, which was proposed by Gaver (1999).

Hummels (1999) illustrated the propensity for interactive-design products to incorporate emotional considerations and suggested that this trend mirrored the growth of technical complexity. A considerable amount of theoretical insight into how goods evoke emotions was provided by Cupchik (2004) and how these ideas had assisted designers in improving their designs to increase emotional impact was discussed. He asserted that the meanings associated with cognitive and behavioural processes were key to improving the designs. Users' expectations and knowledge are integrated into the designs through the use of cognitive/behavioural meanings that incorporate structural, functional, and ergonomic elements. They serve as a vital link between the intended purpose and structural aspects of design outputs and the users who must comprehend and utilise them in their daily lives. Users/consumers who understand and employ designs that are simple and obvious are more likely to do so in the future. To discover and quantify the link between product design characteristics and the emotional responses of product consumers, Harada (1999) proposed a research-based design technique known as 'Kansei engineering'. The appraisal theories of Desmet and Hekkert (2002) argued that consumers' evaluations are critical elements in determining if a design outcome elicits emotion and, if so, what emotion is being evoked by the design outcome (Desmet, 2003; Frijda, 1986). When presenting their point of view on emotional reactions to consumer items, they offered an experimental model that divided product emotions into five categories, that is, surprise emotions, instrumental feelings of utility, aesthetic feelings, social feelings, and interest feelings. Each lesson was demonstrated (with an example) to interviewers to deduce their emotional responses, which were then used to illustrate to the class how to manage their emotional changes. The results of the study indicated that a product may evoke a variety of emotional reactions from users; while the process of consumers experiencing emotion as a result of a design outcome is universal, the emotional responses themselves were complicated and individual. Norman (2004) examined the informa-tion exchanges that took place between users and the design outputs, employing a different approach than that used by Desmet and Hekkert. The visceral, behavioural, and reflective levels of cognitive processing on design were identified and classified according to the conditions and reactions. On the visceral level, the user's first per-ception of the design output and their intuitive emotional responses were discussed in detail. The behavioural level referred to the consuming behaviours taken by con-sumers as a result of the emotions elicited by the design results. The reflective level referred to the level of reflection that people had on their consumption. To illustrate how designs were connected with emotion and delivered enjoyable experiences to build emotional ties with their users/consumers, Norman introduced the notion of 'three levels of design'. Moving even further, Choi (2006) investigated how enjoy-able experiences allow designers to largely satisfy the requirements and desires of users/consumers. He concluded that emotional design would come to life as a result of the creation of a strong and positive mental connection (i.e. emotional concerns in design) that would improve the usability of the designs. The results of Hakatie and Ryynänen (2006), who conducted a basic study experiment, confirmed that the 'three stages of cognitive processing' may be connected to the features of different goods. On the visceral and behavioural levels, the product's selection criteria were more obvious than on the reflective level, which indicated that the product's selec-tion criteria were clearer on the visceral and behavioural levels. Emotion design is recognised as having excellent potential for meeting users'/consumers' wants and desires. Since people often draw emotional messages and experiences from visceral and behavioural levels, the product would be more marketable.

Lo (2007) described the emotional design as the process of putting the user's wants and experience first, based on the theories of Desmet and Hekkert (2002) and Norman (2004), and was influenced by Norman's ideas, who emphasised how the function, shape, and usefulness of components would enhance the user experience, and stated a desire to see more of this in the long term. Unlike previous research, Chitturi (2009) asserted that an effective design should give consumers advantages in terms of total consumption, rather than only in terms of individual consumption. In the real world, consumers of design results experience emotional shifts, both during and after con-sumption. The quality of the advantages in the total consumption experience will cause consumers to experience either positive or negative emotions, depending on their perception of the benefits. Both positive and negative emotions would have an impact on consumers' appraisal of the product, as well as their decision-making throughout the subsequent purchase process. As a result, both positive and negative emotional changes in individuals throughout the course of their total consumption would have an impact on their product loyalty (i.e. design outcome).

In relation to the phrase 'emotional design', only a few definitions are available. More studies should be carried out and recommendations made. According to the ideas discussed above, emotional design refers to a product that has the potential to evoke emotional responses from the user. Emotional design is concerned with the requirements and experiences of the user. Other important requirements and prerequisites for emotional design are as follows:

- The method through which consumers experience an emotion as a result of a design outcome is universal, but the emotional responses that users have are complex and unique to individuals.
- Depending on the scenario and reaction, there are three stages of cognitive processing to consider: visceral, behavioural, and reflective.
- In emotional design, the user's wants and expectations are expressed not only via the design's aesthetics, function, shape, and usability but also through the experience they have built up while using the product.
- To succeed in the real world, emotional design has to be digested by the target audience and become widely available.
- In the long term, both positive and negative emotional changes in users as a result of their total intake will have an impact on their loyalty to the product (i.e. design outcome).

Although several research projects have investigated the notion of 'emotional design', a more specific definition of the term has not yet been given or agreed upon. For this reason, the views presented above from many scholars were reviewed and analysed to gain a more thorough grasp of what emotional design might be.

2.3.2 Designer-driven Research

Decision-making Studies in the Design Process

The function of a designer's decision-making in connection to the design process has been identified in various research investigations. Design was first presented as a decision-making process in 1984. Designers develop goals after understanding the issues at hand. They generate several ideas and choose from a variety of factors when doing so. When they make judgements, they use specific data that include practical factors, related knowledge, personal experience of users' requirements, prior designs, and speculation. Designers regulate data processing by deriving solutions, evaluating consistency, comparing, and selecting. Herrmann and Schmitt (1999) argue that designers must manage the process of converting consumer input into a design idea while adhering to schedule and budget limitations. According to Longueville et al. (2003), decision-making procedures have recently included managerial abilities. A variety of management abilities, including time and risk management, are frequently utilised to prescribe and optimise decision-making. For a systematic decision-making process, designers will need to pick the best management talent. Almendra and Christiaans (2009) emphasised that decision-making in a design process is a customer-centric strategy that is based on a precise and current understanding of the target audience. To address challenges, designers should choose the most effective way to gain such an understanding and the necessary administration skills. The necessity of developing designers' management abilities was also emphasised by Scaletsky and Marques (2009). They suggested that designers learn about material allocation and selection, as this is an important part of the design process. Aesthetics and material applications are important items to consider along with usage effectiveness. In light of the foregoing research on user/consumer-driven design, some design and emotion scholars have focused on designer-driven design (i.e. the relationship between designers and design outcomes). Vosburg (1998) claimed that the emotional changes experienced by designers could drive them to think in new directions and therefore enhance the quality of their ideas. Emotional changes can also help designers to differentiate between diverse types of information, which, in turn, can enhance their capacity to choose the most effective strategies to solve problems (i.e. the decision-making process). Some scholars have suggested that the role of designers should be emphasised in design and emotion studies. Tan (1999) stated that the role of the designer includes storytelling (communicating ideas with their designs) rather than simply pleasing users. From this perspective, the goal of design is to use products to elicit emotional responses from consumers to achieve their own satisfaction. Sanders (1999) stressed the importance of the influence of the emotional responses and experiences of users/consumers on the design process. He stated that the feelings and aspirations of users/consumers can inspire more appropriate and successful designs. The feedback generated by users usually contributes useful information for the generation of ideas and advanced testing for design development. Chhibber et al. (2004) adopted Jordan's (2000) 'four pleasure' framework (i.e. physio-pleasure, socio-pleasure, psycho-pleasure, and ideo-pleasure) and investigated whether designers apply their own knowledge in the design process instead of being solely focused on the user's point of view. They suggested that designers should adopt intuitive methods in the design process. Forlizzi, Disalvo, and Hannington (2003) described how emotion is one of the key elements that influence designers. They argued that changes in the external environment, such as social changes and new interactions between people and objects, influence designers' goal setting and their reflective emotional responses to design (i.e. their emotional experience). Similar to Forlizzi, Disalvo, and Hannington (2003), Ben-Peshat (2004) claimed that designers' emotional changes can increase their sensitivity towards social change and cultural issues (i.e. the external environment). This, in turn, would encourage designers to use their professional skills or even adopt more emotional and intuitive methods in the design process. In this case, designers would be more focused on their own knowledge rather than focusing on users' needs or feedback. Hence, in addition to enriching the users' experience, the design process that focuses on designers' own interests and personal perspectives (mostly involving the designer's emotions) would also establish close relationships between the designers and the public (i.e. the users/audience). This would be as effective as the design process that focuses on users' feedback. Therefore, this approach to the design process could also influence the relationship between designers and their design outcomes.

Departing from previous studies, Aken (2005) focused on the management of the design process. According to Aken (2005), designers learn from people as well as products. Because design activities entail numerous human contacts (client and designer, designer and design team), the designer's communication style will influence decision-making. He pointed out that designers who have more experience and a strong emotional investment in their practice are better able to manage the design process. This would enable them to avoid engaging in unmanaged process design, which can lead to problems with coordination and time management. Enayati (2002) claimed that greater communication skills lead to better judgements. Other elements that designers may manage and that influence their decision-making processes include usability, accessibility, and so forth. Thus, all of the elements under designers' control and that motivate them to make comprehensive considerations in their design processes were rationally studied.

The study conducted by Desmet (2008) centred on the impact of emotional alterations on the practice of design. The research revealed that positive emotions can be advantageous in facilitating the design process. The researcher conducted interviews with multiple designers to investigate the impact of various emotions on their design processes. The aim was to gain insight into how designers' experiences shape their approach to design. Desmet's research revealed that various design processes yield distinct design outcomes, which subsequently evoke diverse emotional reactions from users. Based on the conducted interviews, it was determined that favourable emotions play a significant role in enhancing the creative abilities of designers. Designers often strive to maintain positive emotions throughout the design process, as this is believed to facilitate favourable outcomes. Lacey (2009) conducted a study that examined the impact of user/consumer feedback on the practical design process using Norman's (2004) 'three levels of design' concept as a framework. Lacey's proposed 'favourite mug' methodology revealed that user feedback is primarily linked to physical and emotional factors, which can be attributed to the subjective preferences of both designers and users/consumers. Lacey's study revealed that incorporating information regarding the personal preferences and emotional connections of both designers and users can serve as a source of inspiration during the design process. Specifically, the emotional changes experienced by individuals in relation to products were found to be influential in generating new ideas. Lacey postulated that Norman's three levels of design encompass a reflective level that involves a degree of engagement, which has the potential to foster significant interaction between designers and users. The potential to inform the design of physical solutions can be achieved through engagement at a reflective level in design and emotion theory, ethnographic research, and studio practice.

Studies on Designers and Design Outcomes

The research on designer-driven studies in design and emotion has been influenced by consumer-driven studies on design and emotion (i.e. the relationship between designers and the design outcomes). The quality of designers' ideas will improve if they were more sensitive to emotional shifts, according to Vosburg (1998). To pick the most efficient solutions, designers use emotional shifts to distinguish between different information sources (i.e. decision-making processes). In design and emotion research, designers' roles have been postulated by certain scholars. Instead of satisfying users, Tan (1999) argued that designers should be able to convey stories. Designers should utilise goods to affect consumers' emotions and increase their pleasure. Affective responses and customer experiences have a significant impact on the design process, according to Sanders (1999). He argued that users'/consumers' sentiments and desires may inspire effective ideas. User feedback is typically valuable for generating new ideas and exposing them to additional testing. Physio-pleasure, socio-pleasure, psycho-pleasure, and ideo-pleasure are the four pleasures identified by Jordan (2000). These experts have advised designers to use intuitive techniques. The emotional shifts of designers will assist them to be attentive to changes in social and cultural concerns (i.e. the external environment), according to Forlizzi, Disalvo, and Hannington (2003). This is when professional skills or even more emotive and intuitive approaches could be employed. Designers can thus be more open to changing design approaches, structures, and functionalities. In this way, designers and the public (i.e. users/audience) will be able to establish closer relationships than before. The design process techniques, as well as the interactions between designers and the design outputs, impact the external world. For example, Aken (2005) focused on the management element of design. Designers with great expertise and strong emotions would have superior design management. Unmanaged process design can cause coordination and timing issues. Pleasant feelings are useful in the creative process, according to Desmet (2008). He interviewed designers to learn how different emotions affected their design processes. Desmet discovered through the interviews that different design approaches resulted in variable emotional responses from users. Via the interviews, he also discovered that happy emotions helped designers become more creative. To obtain a successful output, most designers retain positive feelings in the design process. Lacey (2009) investigated how user/consumer input affected the realistic design process using Norman's (2004) three stages of cognitive processing. Lacey's 'favourite mug' research found that empathy is given on both physical and emotional grounds, and this was attributed to the tastes and preferences of both the designers and users/consumers themselves. How the feedback was influenced by users' own preferences and emotional attachment (emotional shifts) was shown. Lacey's results indicate that the reflecting level (in Norman's three levels of cognitive processing) may foster meaningful connections between users and designers. Creating an emotion theory, ethnographic research, and studio practice may help with the design of physical solutions by including the reflective level. Drawing from the preceding research, it is possible to make generalisations regarding the criteria and conditions pertaining to the associations between designers and design outcomes. Designers' emotions are influenced by alterations in the external environment.

The emotional fluctuations of designers have a significant impact on the design process and the resultant structure and functionality of designs. The incorporation of personal experiences and emotions in the design process can enhance its management. The inclusion of emotions in design outcomes, encompassing both material and visual expressions, can foster strong bonds between designers and users. It is observed that designers nowadays tend to rely more on emotional and intuitive approaches to the design process compared with the past.

Studies on Designer Emotions and Decision-making in the Design Process

Several research investigations have established the importance of a designer's personal emotion in the design process. Several studies have been conducted to determine if designers incorporate their own expertise into the design concept rather than focusing exclusively on the user's perspective. Several academic inquiries have investigated the correlation between the emotional states of designers and their creative processes. Several research studies have explored the possibility of designers utilising their personal emotional reactions in the process of designing rather than exclusively prioritising the users' perspectives. Forlizzi, Disalvo, and Hannington (2003) expounded on the impact of emotion on designers by drawing from established theories on emotion and experience. They posited that emotion is a crucial factor that shapes the work of designers. The authors posited that alterations in the surrounding milieu, including societal shifts and novel human–object interactions, impact designers' objective establishment, and their introspective affective reactions towards design, commonly referred to as their emotional experience. Based on the approaches to emotion and experience advanced by researchers in various areas, such as Dewey (1934) and Csikszentmihalyi (1996), the aforementioned studies synthesised previous ideas and demonstrated how the emotional component impacts designers. They thought that changes in the external world, such as societal changes and interactions between people and things, would affect designers' goals and reflect emotional responses (i.e. their emotional experience). These reflected emotional responses would affect their design decisions. However, little study has been conducted on the true links between emotion and designers' decision-making during the design process.

The reflective emotional responses of designers are expected to have an impact on their decision-making processes throughout the design phase. However, scant research has delved into the factual correlation between affect and the decision-making of designers during the design process. Drawing from the aforementioned research, it is possible to make generalisations regarding the criteria and conditions that pertain to the connections between designers and design outcomes. Designers' emotions are influenced by alterations in the external environment.

The emotional fluctuations of designers have a significant impact on the design process and the resultant structure and functionality of their designs. The incorporation of personal experiences and emotions in the design process can enhance its management. The inclusion of emotions in design outcomes, encompassing material and visual expressions, can foster strong connections between designers and users. It is observed that designers presently rely more on emotional and intuitive approaches in the design process compared with earlier times.

Using the E-Wheel Model to Demonstrate the Emotional Functions of External and Internal Design Process Factors

Early research failed to uncover much evidence for the influence of emotion on the design process. Recent research on the design process by Ho and Siu (2009a), however, revealed the function of emotions and proposed the E-Wheel model (Figure 2.1) to explain the relationship between designers, emotions, internal factors (information processing, material allocation, etc.) and external factors (i.e. those uncontrollable by designers, such as technological, social, cultural, and economic factors) of the design process. According to Scherer (1984), emotion is both a pattern of reactions to external stimuli and a process through which stimuli and one's position are evaluated. When designers include their emotional concerns in the decision-making process, their decision-making skill is impacted. Ho and Siu (2009a) investigated how these internal and external elements would impact the entire design process, which includes several decision-making processes and their interrelationships. External variables would alter the designers' emotions, causing them to make new choices that affect those internal elements, affecting and finally changing the design process. This idea elucidates how designers may channel their emotions to generate appropriate responses to optimise their design process and achieve the desired design output.

Based on the aforementioned analysis of the theories surrounding design and emotion, initial research efforts did not yield substantial findings regarding the impact of emotion on the design process. Hence, it is imperative to delve into a more comprehensive analysis of the role of emotion in the process of design. The design process comprises multiple stages. Internal factors, as noted by Cupchik (2004), such as information processing and material allocation, entail diverse forms of decision-making. Designers engage in decision-making processes that involve their emotional responses, which, in turn, impact the internal factors of the design process (Aken, 2005; Vosburg, 1998). The emotional state of a designer has the potential to impact internal factors, leading to varied design outcomes. Designers' decision-making can be influenced by various external factors that do not involve their emotions (Aken, 2005; Almendra and Christiaans, 2009). The decisions made during the design process have an impact on internal factors, which, in turn, influence the final design outcome.

The emotions of designers may be influenced by external factors in the design process, which are not directly under their control. This statement aligns with the perspective of Forlizzi, Disalvo, and Hannington (2003), which suggested that alterations in the surrounding context may impact the introspective and affective reactions of designers (i.e. emotional experiences). The impact of external factors on the emotions of designers is primarily contingent upon the degree of awareness and introspection that a designer possesses with respect to the external milieu (i.e. the external factors of the design process). The incorporation of emotional considerations into the design process may potentially augment the decision-making capacity of designers. Furthermore, the subsequent emotional changes experienced by designers serve as a motivating factor that prompts them to explore different avenues of thought and ultimately leads to an improvement in the quality of their ideas (Vosburg, 1998). Designers may opt to approach their design challenges from multiple angles after improving the calibre of their concepts. Incorporating emotional considerations into the design process can enhance the decision-making capabilities of designers. The incorporation of emotional changes may aid designers in distinguishing between various forms of information, thereby facilitating the selection of optimal problem-solving techniques (i.e. decision-making processes). The emotions of designers are influenced by modifications made to external factors during the design process.

The affective reactions of designers have an impact on their decision-making processes. The extent to which external factors impact the emotions of designers is primarily contingent upon the level of awareness and reflection that designers possess with regard to their external surroundings. Alterations to a decision have an impact on the internal factors, which are the factors that are within the control of the designer throughout the design process. Consequently, the modifications made to the internal factors lead to changes in the decisions taken in the subsequent stages of the design process. According to Kaufmann's (2003) assertion, certain designers believe that the presence of affirmative emotions may facilitate the cognitive processing of information. The individual's capacity to scrutinise incoming data is heightened, thereby augmenting their aptitude for making informed decisions. The incorporation of emotions into the design process is known to augment the overall design process, as information processing is an intrinsic factor in the said process. As a result, the design process is optimised to achieve the desired design outcomes. According to Best (2006), the inclusion of emotion in the design process is crucial to ensure the quality of the resulting design outcomes. Consequently, the design results are impacted.

Emotional Functions in Relation to External and Internal Factors

On the one hand, internal variables include a variety of decision-making processes. Designers will then make appropriate decisions (using their emotions) to impact the design process's internal variables (information processing, material allocation, and so on). As a result, emotion may have an effect on the internal elements that contribute to the variations in design outputs. On the other hand, the external variables will have a direct effect on how designers make their own judgements (without their emotional engagement). As a result of the decisions taken, the internal aspects of the design process are impacted, and as a result, the design product is impacted as well.

Framework for the New Model (the E-Wheel Model)

Ho (2014) examined the role of emotions in the design process and introduced the E-Wheel model (Figure 2.1) in his paper, 'The New Relationship between Emotion and the Design Process for Designers' (p. 52) as a means of elucidating the interplay between designers, emotions, internal factors (such as information processing and material allocation), and external factors (such as technological, social, cultural, and economic factors). This model was developed based on research into the design process. As per Scherer's (1984) perspective, emotion is characterised by a set of responses that are elicited by external stimuli and lead to the formation of appraisals regarding both the stimuli and the individual's circumstances. The incorporation of emotional considerations into the decision-making process has a significant impact on the decision-making capabilities of designers. The impact of internal and external factors on the design process was investigated by Ho (2010), wherein various decision-making processes were involved. The emotions of designers can be influenced by external factors, resulting in altered decision-making that impacts internal factors and ultimately affects the overall design process. This procedure offers a glimpse into the ways in which designers can leverage their emotions to formulate suitable reactions that enhance the efficacy of their designs.

As depicted in the E-Wheel model above, external influences in the design process (i.e. those that are outside the designers' control) will influence designers' emotions. This is consistent with Forlizzi, Disalvo, and Hannington (2003) who asserted that changes in the external environment impact designers' reflective emotional responses (i.e. emotional experience). The extent to which external variables influence designers' moods is mostly determined by designers' knowledge of and

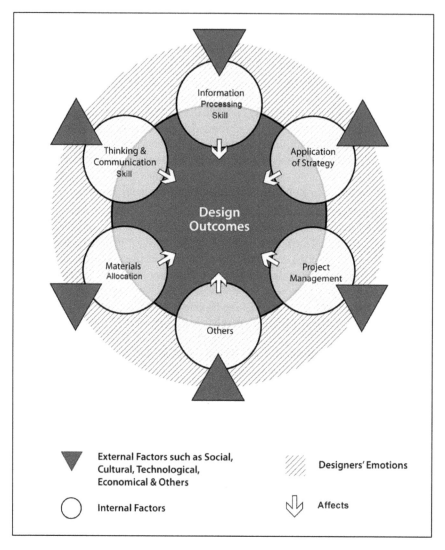

FIGURE 2.1 E-Wheel model

contemplation of the external world (i.e. the external factors of the design process). When designers add their emotional concerns to the process, their decision-making capacity is strengthened. Designers' emotional shifts will cause them to think in novel directions, thereby improving the quality of their ideas (Vosburg, 1998). Emotional shifts also assist designers in differentiating disparate pieces of information, allowing them to pick the most effective ways to resolve challenges (i.e. decision-making processes).

In contrast, internal factors (those under the designer's control) comprise a variety of decision-making processes. When designers make decisions based on their emotions, the internal elements of the design process (information processing, material allocation, etc.) are impacted as well. According to Kaufmann (2003), some designers have also discovered that pleasant emotions can aid in information processing. Their capacity to analyse received information will improve, along with their ability to make judgements based on that knowledge. Because information processing is one of the internal components in the design process, including emotions will benefit the entire design process. As a result, the related design process will optimise the output of the design. As Best (2006) noted, an emotional component to the design process will ensure the quality of the design outputs.

A Case Study – An Application of the E-Wheel Model

The E-Wheel model's proposed concept that explains the relationships between external factors affecting designers' emotions (for appropriate responses) and those emotional changes impacting their decision-making process (across the entire design process) can be explained and illustrated further by the following case studies. Tommy Li infused his Honeymoon Dessert rebranding campaign with a visual language grounded in modern Orientalism. He employed a selection of quaint, 'sweet', and nostalgic pictures that graced the dinnerware, menus, and takeout goods.

To capitalise on the nostalgic movement among the young of Hong Kong's population, designer Tommy Li developed a visual language rooted in modern Orientalism for his Honeymoon Dessert rebranding campaign in 2006. He first reviewed the concept with the company's six owner-shareholders and then examined the tendencies of Hong Kong's young people before settling on the nostalgic concept. Tommy Li and his crew began collecting photos from vintage Hong Kong publications and posters that were appealingly quaint and 'sweet'. For instance, they saw depictions such as 'A young girl grins as she thinks about her boyfriend' or 'A plump infant swings his arm joyfully in the air'. The decision to select a particular image was made mostly on the basis of how it evoked an emotional reaction (including a subjective sense of nostalgia) in design team members. When the designers viewed those ancient Hong Kong images with symbolic value, they developed a feeling of identity. The design team inserted their individual interests, intuitive evaluation, and humour into the process of designing promotional products and interior décor. These vintage Hong Kong photographs served as powerful emotional triggers for society. Simultaneously, the team created several 'monster' characters largely based on the characteristics of the company's six owner-shareholders as viewed through the designers' eyes. These unique cartoon-like figures enhanced the design team's humour and were well received by local customers.

Chau Carrie Wun Ying is the primary illustrator at Homeless, a Hong Kong lifestyle concept shop. She is also the creator of the Wun Ying Collection, which incorporates her unique illustrations into the design series' manufacturing line. Her celebrity continues to rise as a result of her creative characters and new goods. Through her well-known pictures, she conveys positive messages of pleasure and love through her artwork. According to Chau (2007), she is mostly inspired by the 'abnormal' objects that she finds around her as a starting point for her creative approach. These might range from a print, a narrative, people, the environment, and current events to significant events that have occurred in her immediate vicinity and impacted her emotions. Rather than paying close attention to consumers' perspectives, she incorporates her own emotions into the design process. For instance, in the process of developing each matchbox in her collector series, 'Spark', Chau incorporated her emotional responses to 'those single sparks throughout her drawing career' into the decision-making process as she began to determine which technique or mode of expression she would employ while incorporating her professional knowledge and skills into the design process. 'Spark', a matchbox collector series, preserved a succession of brief yet priceless experiences throughout Carrie Chau's career. In this example, the designer's emotional reactions to external variables (i.e. narratives, people, the environment, current events, personal interests, and intuitive judgement) influenced her decision-making, which was all included in her design process.

These two case studies eloquently corroborate the suggested conceptual E-Wheel model, which states that external variables impact the designer's emotions, which in turn affect the many decision-making processes involved in the design process's internal components. As a result, the altered design process will result in altered design results. Thus, the E-Wheel model elucidates the notion that the designer's emotions will become the primary and guiding element in the design process and its functions in reaction to external and internal design elements. As noted previously, a number of criteria and circumstances concerning the link with designers and design outputs have been established:

- Changes in the external environment have an impact on the emotions of designers.
- In addition to having an impact on design processes, designers' emotional shifts also influence the structural features of the design outputs.
- Designers' ability to control the design process would be enhanced if they included more personal knowledge and intense emotion into the design process.
- A tight link between designers and users may be achieved through design outputs (including both physical and aesthetic expressions) that incorporate emotional components.
- Designers increasingly rely on emotive and intuitive techniques in the design process as opposed to previous generations.

Interactions between users/consumers, designers and outcomes besides users/consumers-driven studies and designer-driven studies, additional research has emphasised the communication between users/consumers and designers through design results or, to put it another way, the interaction between users/consumers, designers, and design outcomes. Funke (1999) investigated how emotions interact with goods (i.e. design outputs) to meet subjective expectations and suggested that design functioned as a semiotic tool in an experience market to achieve this goal. His research revealed that consumers' emotional concerns might be converted into a product (i.e. design results) that, in terms of its function, met their personal expectations and enhanced their overall experience. As a result, consideration of emotional factors in the usage and design process should take precedence over the evaluation of functional aspects. Jordan (2000) investigated the interaction between people (including designers and users/consumers) and goods from a holistic perspective as well as the criteria for evaluating the quality of designs. The designer's point of view was that it would be sensible and reasonable to listen to the consumers' demands and then create the product with skill and judgement, empowering the people while also providing them with pleasure. When people utilise the product, they would experience emotional advantages, since the product would have an effect on their state of mind. Suri (2003) investigated how the design of a product impacts the customer's experience. She discovered that emotional concerns had an impact on both designers and users when it came to developing and consuming design results, based on her study on 'design expression' (i.e. designer-driven studies) and 'user experience' (i.e. user-driven studies). Because products (or design outcomes) are increasingly similar in many aspects and attributes, including technology functionality, price, and quality, designers are being asked to create more differentiated outcomes to meet the needs of users/consumers, which is becoming increasingly difficult (Suri, 2003). As a result, designers are encouraged to use their design outputs to influence people's actions and perceptions. Emotional considerations have an impact on the approaches taken throughout both the design process and the consumption of the final product. They are also an essential component that motivates designers to investigate design concepts that add to the user's experience and conveys experiential ideas to the public (between the users and designers). Cupchik (2004) developed the notion of 'creating for experience', which is based on a concept similar to that of Suri, to explain the interactions that exist between designers, design outputs, and consumers. During the process of experience design, the designer assigned meanings or messages to the design object. The intended purpose of the design was utilised by the users/consumers, who were subsequently affected by the designer's intended creative statement. According to the information presented above, designing with emotions appears to have the potential to meet both what users/consumers anticipate and what designers can plan and provide as part of the interactive experience. Tzvetanova (2007) proposed notions that were similar in that they were concerned with understanding the link between the emotional responses of the customers and the designers. Tzvetanova emphasised the importance of the interactive connection that exists between users and design professionals. The user factors (i.e. the feedback from the users) were the most significant element that determined the overall quality of the emotion design project. After all, the number of design studies that have expressly incorporated the phrase 'emotion design' with a clear and comparably precise meaning is extremely limited in comparison to the overall number of studies. It is only feasible to summarise the fundamental notion of 'emotion design' suggested by the many academics listed above; nevertheless, some essential requirements and prerequisites for emotion design are as follows:

- Design serves as a semiotic instrument and conveys the messages of the designer to the target audience in most cases.
- The planned function or design creates an interaction between the users/consumers and the designers. Affective considerations of both users/consumers and designers influence the approaches taken throughout both the design process

and the creation of the final product. The designers first gain an understanding of the consumers' experience before exploring design solutions that would improve it. Eventually, the ideas generated by users/consumers and designers are communicated to one another.

• Emotional concerns influence approaches to both the design process and the consumption of the design; these include understanding the users' experience, exploring design concepts that contribute to the users' experience, and communicating the experiential ideas between users and designers. Understanding the users' experience, exploring design concepts that contribute to the users' experience, and communicating the experiential ideas between users and designers are all affected.

Despite over a decade of progress in the study of emotion and design, there were just a few occurrences of the phrase 'emotion design' in research publications. Scholars prefer to use the term 'emotional design' to define all designs that incorporate or are connected to emotion rather than the term 'emotion design'. Several notions and associated theories presented by various scholars have been examined to gain a better grasp of what emotion design is and how it may be defined more precisely.

According to the preceding theoretical examination of all of the ideas advanced by scholars, it appears that there are certain misconceptions and instances of confusion regarding the application of these words. All of the research papers that have been addressed and analysed above were among a small number of studies that attempted to identify and investigate the fundamental definition, nature, and features of emotion design as well as the terms 'emotional design' and 'emotionalise design' (Ho and Siu, 2009a). The majority of the studies used these words to describe designs that included or were associated with emotion. The result is the development of a new model that has a clear definition and nature in addition to the links between the many concepts that are being discussed.

2.4   A NOVEL APPROACH TO EMOTIONALISE DESIGN, EMOTIONAL DESIGN, AND EMOTION DESIGN

The preceding review of literature has demonstrated that various synonymous terms are employed to denote the concept of design and emotion. The precise delineations of concepts such as 'the affective dimensions of design', 'design that elicits emotional responses', and 'design that incorporates emotional elements' remain somewhat ambiguous. The present section aims to provide a more precise and tangible explication of the term 'design and emotion' in order to achieve an all-encompassing comprehension of the phenomenon.

2.4.1   THE NEW MODEL'S FRAMEWORK

The terms 'emotional element of design', 'emotional design', and 'emotion design' were not well defined in the aforementioned research papers. While emotional expressions perform critical communication roles and have an effect on information processing, the modes of communication may be expressed plainly using terminology such as information encoding and decoding (Wogalter, Dejoy, and Laughery,

1999). The designers encrypt the data, while the users decrypt it. The amount of information digested is determined by the consumer's level of comprehension. Thus, the communication system (information and decoding) may be utilised as a foundation for developing a new model that exemplifies the three essential phrases: emotionalise design, emotional design, and emotion design.

2.4.2 Proposed New Model

To begin, it is critical to clarify the function of emotion in the usual design cycle before proposing a new model to describe the link between these concepts. According to the overview of theory above, emotion should be divided into three distinct components throughout the design process: designers, design output, and users/consumers. As a result, this will serve as the foundation for our suggested model. While the majority of theories presented by researchers such as Desmet (2002) and Norman (2004) place a premium on the link between the design output and users/consumers, it has been claimed that emotion is a significant influence in a designer's decision-making process.

As a result, a new 3E model (Ho and Siu, 2012) may be offered as a framework that explains the tight interactions between designers, design outcomes, and users under the umbrella of emotion and design while also explaining the philosophy and principle behind it. This 3E model (Figure 2.2), in *Emotionalise Design, Emotional*

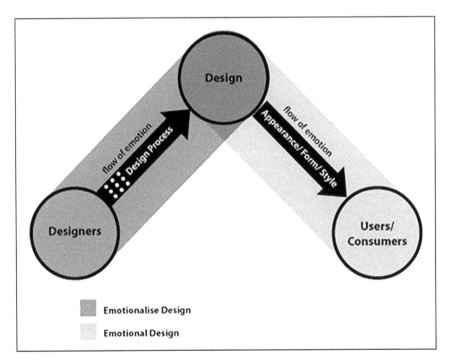

FIGURE 2.2 Model to demonstrate the link between 'emotionalise design' and 'emotional design'

Design, Emotion Design: A New Perspective to Understand Their Relationships (p. 20), is devoted to elucidating the tight link between emotionalise design, emotional design, and emotion design. These three critical core aspects – designers, design outputs, and users – are always incorporated into the standard design cycle and serve as an essential foundation.

2.4.3   EMOTIONALISE DESIGN

Prior to the final design output being released to the market, designers must go through the design process. Frequently, designers include or infuse numerous emotional considerations into the design process to arrive at the desired design output. From another angle, the output of the design reflects some aspects of the designer's mood. In this circumstance, designers inject their own emotions into the design process to create a design outcome, which the designers may refer to as 'emotionalising design'.

A typical illustration is the RMX Extended Mix book. Rinzen, an Australian collective of artists, designed the book in 2001. Rinzen introduced RMX Extended Play, an intimate look at the visual design (Alexander, 2001). According to Adrian Clifford, one of the group's members, each member created an initial piece on one of eight topics. The files were then transferred to each designer in turn, being altered, enhanced, and partially deleted along the way. In this scenario, designers could articulate their own opinions on matters of personal interest with professional aplomb. Their emotions, personal interests, and intuitive judgement could thus be incorporated into the design process, although they initially did not view the product through the consumer's eyes.

Another example is the design projects assigned to students. When student designers strive to meet the learning objectives for their projects, they include elements of their own emotions in their work. While the design output may be capable of eliciting emotional responses from users/consumers, it will not be directly consumed by users/consumers in a real market under typical conditions. As a result, design student projects in schools may only be considered examples of emotionalise design.

In fact, identifying examples of emotionalise design is difficult, as designers may incorporate specific types of emotions into the design process to achieve the desired result, and the majority of them are likely to convert into emotional design once they reach the market. Design results may elicit specific emotions to connect with and interact with users/consumers. This encourages them to respond to the design's look or function. For instance, the RMX Extended Mix book was eventually made available for customers to read, which elicited emotional responses. Thus, depending on the level of the design at issue, the book might be classified as an emotional design.

These two instances exemplify the suggested idea of emotional design while also conforming to the conceptual E-Wheel model. Ho (2010) argued in the E-Wheel model that external variables influence designers' emotions, which subsequently influence various decision-making processes inside the design process. Consequently, the altered design process will result in altered design results. Another consequence is that the designer's emotions will become the driving force behind the design process and its functions in reaction to the external and internal variables that affect the whole design process.

With the RMX Extended Mix book, it was discovered that external variables (such as technological progress and cultural background that adds to the subject) impacted both the logical reasoning and emotional responses of designers when working on their unique design themes. Rather than prioritising the consumer's perspective, the Rinzen artists infused the design process with their own feelings. When each designer began designing or editing, they contributed their emotional states to the decision-making process along with their professional knowledge and abilities in the design process. In this scenario, their emotional responses to external circumstances, personal interests, and intuitive appraisals all influenced their decision-making, as all of these elements were incorporated into their design processes.

In the case of design assignments that required students to accomplish learning objectives through their projects, external elements affecting the design process (e.g. the student's cultural background) would impact the inexperienced design students' emotions. They put their emotions into play when making judgements on internal aspects (information processing, material allocation, and so on) and therefore governed their own activities and made distinct design decisions. As a result, the altered design methods would have had a commensurate effect on the design outputs. According to this case study, external variables influenced inexperienced design students' emotions, while their emotions may have influenced their decision-making about internal issues. As a result, the design process and results would be impacted.

This is in contrast to the notion of '[to] emotionalise design', in which designers incorporate their own feelings into the design process to achieve a desired result. It says that designers must complete the design process before the final product is released to the market. Designers frequently include or integrate numerous emotional considerations into the design process to achieve the desired design output. Another angle is that the output of the design reflects some aspects of the designer's personal feelings.

2.4.4   EMOTIONAL DESIGN

When a design outcome is created and placed on the market, it may have the capacity to influence the consumers' emotions when they consume it. This type of emotional characteristic (or motivation) is derived mostly from the design style, function, form, usability, and experience encountered by the consumers. In other words, the design outcome may create a shift in the users' feelings and elicit emotional reactions, such as happiness, irritation, or excitement. In this case, the design outcome may encourage users to link, recollect, or envision similar events or experiences that affect or elicit their own emotions. From the user's perspective, this is referred to as emotional design.

A classic example is the Japanese designer Kenya Hara's signage design at the Umeda Hospital in Tokyo (Mollerup, 2005). Kenya Hara and Yukie Inoue designed the graphics. The signpost is composed of cotton and fabric; it has a pleasant, soft feel that helps to calm the patient's mind and alleviate anxiety and concern. Thus, this signage might elicit an emotional reaction from the user, which is referred to as emotional design.

2.4.5   EMOTION DESIGN

If we examine the entire process of emotion flow, from the designers who inject their emotions into the outcome (emotionalise design) to the users who are moved to specific emotional responses as a result of consuming the design outcome (emotional design), we see interactions between the designers and the users via the design outcome, thereby establishing a strong relationship between these three roles. This also serves as a foundation for emotion design. Typically, in emotion design, design reflection occurs following design consumption. Designers can get direct input from consumers (in the form of post-consumer comments, for example), or users might provide indirect feedback to designers.

The example of Emoticons from Icons-Land (2006) illustrates the notion of 'emotion design' through case studies. This is a common use of 'emoticons'. On the Microsoft Network (MSN), emoticons are used in electronic conversations between friends to convey the sender's emotions: 'Emoticons are emotional graphics – visual representations of how you feel when words alone are insufficient' (MSN, n.d.). MSN, one of the leading creators of emoticons, has declared unequivocally that these symbols elicit emotions in consumers. To symbolise their own emotions, users can use any of these symbols depending on their look, shape, and style. This is referred to as emotional design in which the design outputs (emoticons) are imbued with certain emotions to elicit a response from the user. However, the design approach for these icons is also intended to 'emotionalise' design. Because the icons are created in response to the designer's feelings, the design process incorporates emotion. Both emotionalise design and emotional design exist in this scenario, and the designers may engage with the consumers through the emotion flow, making this a typical example of emotion design.

2.4.6   VISCERAL LEVEL, BEHAVIOURAL LEVEL, AND REFLECTIVE LEVEL

Drawing from the preceding analysis of the theory of emotion design, it can be posited that the function of emotion within the broadest design framework is comprised of three primary constituents, namely designers, design outputs, and users/patrons. The process of encoding information by designers and decoding it by consumers is complemented by the emotional expressions of both parties, which are developed based on emotional concerns and serve a crucial communicative purpose.

The process of designing involves a procedure of 'information encoding and decoding' (Wogalter, Dejoy, and Laughery, 1999) that aims to recognise the interactive connections among three key roles, namely designers, design outcomes, and users/consumers. Scholars, such as Desmet (2002) and Norman (2004), have directed their attention towards examining the correlation between design outcomes and users/consumers. As per scholarly research on the design process, it has been suggested that the design process comprises three primary roles, namely designers (Tan, 1999; Vosburg, 1998), design outcomes, and users/consumers (Jardon, 2000). In order to examine the impact of emotion on the design process, it is imperative to thoroughly consider the connections between emotion and design outcomes as well as the relationships between users and designers. This study proposes three novel categories of design and emotion research, namely user/consumer-driven research, designer-driven research, and research on the correlation between users/consumers and designers as manifested in design outcomes. According to this classification approach, the integration of emotion in the design process entails three distinct roles, namely designers, design outcomes, and users/consumers. The integration of emotion into the design process should be a paramount consideration at every stage. The inter-dependent connections between emotion and the three roles have been formulated into three fundamental concepts, namely emotionalised design, emotional design, and emotion design.

It is thus possible to describe the notion of 'emotion design' by referring to three levels of processing: the visceral level, the behavioural level, and the reflective level, as stated by Norman (2004). As seen in Figure 2.3, Ho and Siu (2012) proposed, in *Emotionalise Design, Emotional Design, Emotion Design: A New Perspective to Understand Their Relationships* (p. 21), that the first level of emotion design is primarily concerned with the visceral level. A consumer's brain receives signals at

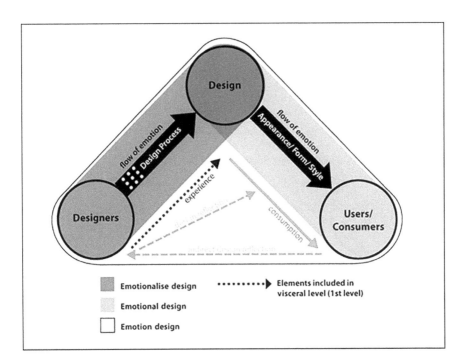

FIGURE 2.3   3E model illustrating the visceral level

this level through their first perception of a design output (which was derived from the designers' design experience) and responds instinctively by eliciting emotional reactions from the brain.

The behavioural level (Figure 2.4), as illustrated by Ho and Siu (2012) in *Emotionalise Design, Emotional Design, Emotion Design: A New Perspective to Understand Their Relationships* (p. 22) is the second level of emotion design, and it is where the designers' design experience and users' consuming activities come together to benefit both the designers and the consumers. Regarding the design process, designers infuse their emotions to achieve a desired design output, whereas consumers consume the design to experience the emotions produced by the design product.

At the third stage of emotion design, known as the reflective level, participants reflect on their previous experiences. This has three results: 1) In the course of their everyday work, designers receive feedback and indirect reflections from their clients. 2) Thus, the designers may be able to gain more insight into the evolution of their design in the future. 3) Consequently, both direct and indirect reflections may be seen in this third level of emotion design, which is also known as the emotion design level (Figure 2.4).

Following the ideas presented by the 3E model, designers incorporate specific types of emotions into the design process to achieve the design results, a process known as the emotionalisation of design (Figure 2.5), as illustrated by Ho and Siu (2012) in *Emotionalise Design, Emotional Design, Emotion Design: A New Perspective to Understand Their Relationships* (p. 22). Taking an alternate view, the design results

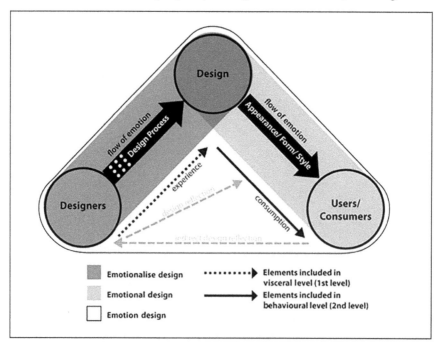

FIGURE 2.4   3E model illustrating the behavioural level

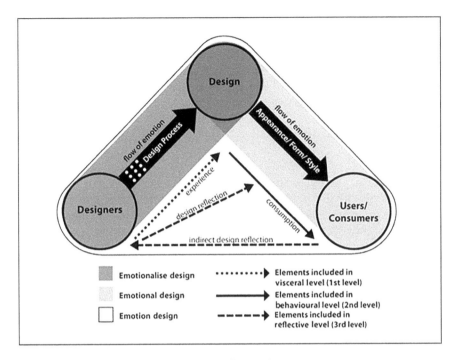

FIGURE 2.5 3E model illustrating the reflective level

contain specific types of emotions that are injected into the design process by the designers in order for them to be accepted by the target audience. As a result, emotion may come to play a significant part in the design process, which may in turn have an impact on the emotional reactions of the users. According to the results, this represents a new opportunity to understand how emotions, in addition to well-known factors such as ergonomics (Wickens and Hollands, 2000), sustainability (McLennan, 2004), and accessibility (Mace et al., 1996), could be a significant consideration in the design process going forward. Aspects of the design process that must be understood include how emotions function in the design process and to what extent they influence designers or design results throughout the design process.

The preceding literature review examines diverse concepts and theories pertaining to the intersection of design and emotion, encompassing the proposed E-Wheel and 3E frameworks. The evolution of the design and emotion literature highlighted significant concepts and theories. The material presented herein offers a comprehensive overview of the evolution of the design and emotion paradigm, tracing its roots from the inception of emotional theory to its cross-disciplinary applications and, ultimately, its impact on the field of design research.

The introductory section of this chapter outlines the objective of the literature review, which is to provide an in-depth historical analysis of the design and emotion concept. The ultimate goal is to contribute to the existing body of knowledge on this subject matter. The literature review's discoveries offer a theoretical comprehension of the ways in which emotions impact the process of design. This study places emphasis on the impact of emotions from both the user's standpoint and the designer's decision-making during the design phase. The E-Wheel and 3E models elucidate the interconnections between emotions, decision-making, and the design process. The present study provides clarification on the key terms, namely emotionalised design, emotional design, and emotion design. Empirical Studies 1 and 2 utilised the aforementioned concepts and terms to examine the comprehension of the design and emotion concept among novice design students.

2.5   SUMMARY

As indicated in the extant literature, the majority of studies pertaining to design and emotion centre on the affective reactions of design product users. In recent times, the scope of this particular field of study has broadened to encompass variables pertaining to design results such as 'affective functions' and 'experience design'. In contrast to prior research that solely concentrated on users' reactions to designs, certain design experts have delved into the roles of emotions in design processes. The literature suggests that there is a correlation between the emotional responses and decision-making of designers and the design process. Moreover, the existing theoretical literature offers substantiation that emotions hold significant importance in the process of decision-making. Several theories have been formulated regarding the correlation between the emotional responses of designers and the design process. These include the E-Wheel model and the 3E model among others. The applicability of these theories should extend beyond the design research domain and experienced designers to encompass design novices. Notwithstanding this, as indicated in the first chapter, inexperienced design students encounter challenges in accomplishing their tasks and may encounter difficulties in managing the design procedure. The students appear to possess inadequate comprehension regarding the correlation between emotion and the process of design. This prompts inquiry into the sufficiency of emphasis on the subject matter of design and emotion within contemporary design pedagogy. Is there an awareness among students regarding the potential influence of concealed variables that might impede their ability to effectively participate in the design process? Could the incorporation of instruction on design and emotion serve as a viable strategy for facilitating the design process for inexperienced students? This research study aims to investigate the general understanding and perception of design and emotion among novice design students. Additionally, it seeks to explore the relationship between emotion and the decision-making processes of these students during their design work.

Comprehending the correlation between emotion and the design processes of novice design students would facilitate their comprehension of the significance of emotion in their design studies. Comprehending the aforementioned concept would enable individuals to effectively recognise and regulate their own emotional states, thereby facilitating superior decision-making abilities and optimal design results, irrespective of whether they experience positive, negative, or neutral emotions. The findings of this investigation offer valuable perspectives for instructors of design with respect to the correlation between design and emotion and the design methodology of inexperienced design students. Through an examination of this correlation, the present research will offer a significant contribution to the domain of pedagogy in design.

The creation of this conceptual perspective is meant to improve the knowledge of the interaction linkages between the levels to create a more complete and trustworthy description for these three important concepts that frequently emerge in previous research in the area of design and emotion. When describing design that is emotional, writers frequently refer to it as emotional design, which is derived from the English grammatical usage. Emotional design can be defined as design that carries emotion in some way or design that can elicit or encourage some emotions in the users through form and appearance. The phrase has, nonetheless, become a general keyword for everything that involves emotion in the majority of research, yet only a small number of studies have looked into the precise definition of the term and the evidence that supports it. As guidance on how to accept these three essential phrases with more specific meanings, the suggested model is beneficial by introducing a new theoretical notion. Making clear distinctions between the phrases might also be beneficial by helping to avoid any future misunderstandings about the subject. In order to make the concepts more practical, further case studies with applications might be offered and presented in a future publication.

The present literature review has furnished a historical overview of the concepts of design and emotion. In recent decades, several theories have been posited regarding the correlation between design and emotion as well as elucidating the connection between design and emotion. The study of design and emotion offers a conceptual framework for comprehending the impact of emotions on design. This can be distilled into the following: A multitude of scholars have examined and scrutinised the utilisation of terminology pertaining to design and emotion, including emotion design, emotional design, and emotionalised design. Nonetheless, a limited number of scholars have effectively explicated the essence and attributes of these specific concepts.

In the realm of academia, it is common for scholars to employ the phrase 'emotional design' in accordance with the conventions of English grammar. This phrase is utilised to denote a design that possesses an emotional quality, either through its incorporation of emotions or through its ability to elicit emotional responses from its users based on its form and appearance. The majority of extant scholarly literature explores the correlations among end-users or consumers, design outputs, and affective responses as opposed to the design procedure itself. The design process involves considerations of economic factors.

The 3E model, which has been proposed, is a novel theoretical framework that provides a relatively precise interpretation of the aforementioned terms. The model distinguishes between the three terms in order to enhance the precision of the subject matter. The 3E model was formulated to enhance the theoretical comprehension of the interdependent associations among the concepts. This facilitated the development of a more encompassing definition of the three fundamental terms than the one currently present in design and emotion research. As per the 3E model, 'emotionalised design' refers to the incorporation of the designer's emotions into the design process.

Emotional design is perceived as a type of design that has the ability to elicit emotional responses from users or consumers. Emotion design pertains to the incorporation of emotional considerations in the designer–user interaction within the design process. The following chapters provide case studies and practical examples to demonstrate the implementation of the 3E model in real-world scenarios.

3 Research Objectives

3.1 AIMS AND OBJECTIVES

In this study project, both exploratory and descriptive research methodologies were employed. Quantitative data can offer a framework for qualitative data while a qualitative approach can compensate for the limitations of quantitative data by exploring the possible interpretations (Creswell, 2008). As a consequence, the function of emotion and its relevance in the creative processes of inexperienced design students can be grasped from many viewpoints. Through comparison research that utilises both quantitative and qualitative techniques, one can examine how emotional shifts might substantially impact such students' decision-making and design processes.

The objective of this study is to gain a general grasp of the ideas of design and emotion of designers in the current generation. It studies the function of emotions and their relevance to decision-making and design processes. It also helps to understand why the designers of the current generation have difficulty in controlling design processes.

The study focuses on three primary areas of theory that are its key objectives: 1) to explore the general understanding of the concepts of design and emotion of designers of the new generation, 2) to investigate the role of emotions and their importance in designers' decision-making and design processes, and 3) to explain why designers of the new generation have difficulties handling design processes.

3.2 RESEARCH SCOPE AND QUESTIONS

The following are the research questions for this study: Is there a link between emotion, decision-making, and the design process? Is there a relationship between emotion, decision-making, and the design process? If so, what exactly is it? What is the understanding of the idea of design and emotion among students? The following should be investigated further in future research:

DOI: 10.1201/9781003388920-3

1. How do designers of the new generation understand the link between design and emotion? And how do they understand the relationship between design and emotion?
2. What is the significance of emotion in the design process and what function does it possess?
3. In what ways does emotion influence their decision-making and design processes?

4 Research Method

4.1 RESEARCH DESIGN

The entire study was divided into two sections: Empirical Study 1 and Empirical Study 2 (Figure 4.1). Empirical Study 1 examined the overall perceptions of inexperienced design students regarding the ideas of design and emotion. Empirical Study 2 refined the results of the first empirical study, analysed why students struggle with design processes and examined the function of emotion and its significance in the design processes of inexperienced design students. Both studies used distinct research methods and analyses to arrive at their conclusions. For Empirical Study 1, a questionnaire (i.e. quantitative research) was utilised to ascertain inexperienced design students' general comprehension of the concepts and terms associated with the theme of design and emotion. In Empirical Study 2, a focus group with a semi-structured interview (i.e. qualitative research) was utilised to analyse inexperienced design students' experiences and understandings of design and emotion and their issues with design processes. The following paragraphs will discuss the methods used for each study section.

4.2 EMPIRICAL STUDY 1: QUANTITATIVE RESEARCH

Quantitative Research was Based on a Questionnaire to Investigate Design Students' Overall Comprehension of Design and Emotion

4.2.1 PARTICIPANTS

The questionnaire was used to ascertain designers' overall impressions of three synonymous terms: 'emotion design', 'emotional design', and 'emotionalise design' as well as their knowledge of and experience with design and emotion (Figure 4.2). The research entailed enrolling 120 inexperienced design students from Hong Kong's degree programmes. Students who had studied design for a total of two years (one year for bachelor's degree students and two years for sub-degree students) were targeted according to their talents and training. This set of students was picked to ensure they had not completely established their own design thinking and had no prior experience managing design projects autonomously. Because the students were new to design

DOI: 10.1201/9781003388920-4

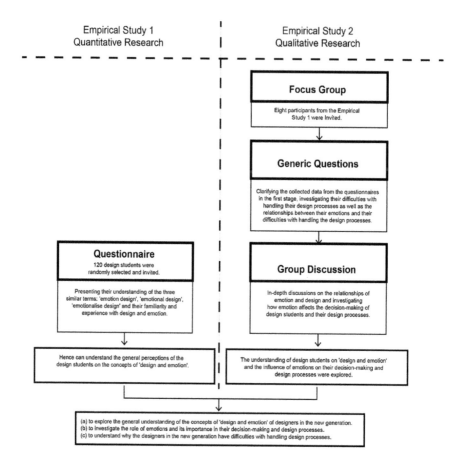

Research framework for the whole study

FIGURE 4.1   A research framework

expertise, it was assumed that the collected data would be unaffected by potential confounding variables, such as skill and knowledge transfer via instructors and reference books. Additionally, the students acquired fundamental ideas of the design process, such as design thinking, design methodology, the role of designers and design outcomes, and the role of customers, all of which fall under the umbrella of design study, making the students excellent candidates for the study. It was believed that the pupils would experience only moderate emotion and refrain from becoming overly enthusiastic or anxious, which could potentially impact the outcome. A total of 120 students were randomly selected: 68 (56.7%) were female, while 52 (43.3%) were male. According to Lodico, Spaulding, and Voegtle (2010), data collection helps researchers with limited resources and time to acquire a broad picture of a population. As a result, the 120 participants allowed for generalisation of the study's results to the full group of design students.

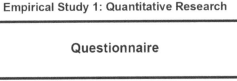

FIGURE 4.2   The quantitative research structure of Empirical Study 1

In most studies of emotion, gender difference is a factor that may impact the outcome. Numerous research efforts have studied gender differences in a range of emotional phenomena, including recall for various emotional episodes (Seidlitz and Diener, 1998), non-verbal emotion communication (Buck, Miller, and Caul, 1974; Wagner, Buck, and Winterbotham, 1993), and assessments of non-verbal cues (Hall, Carter, and Horgan, 2000). Females exhibit greater emotional reactivity than males. They describe their experiences more vividly than males do and exhibit more emotional responses; for example, ladies smile more than men do when recalling moments of happiness and love. The difference between males' and females' real emotional responses also has an effect on their performance in the organisation and the way their emotional shifts are described (Bradley et al., 2001; Schimmack, Oishi, and Diener, 2002). Given the distinct requirements of males and females, a set of possible factors was established for each question to enable participants to rank their views of design and emotion. To elicit participants' descriptions and expressions of their thoughts and impressions, all possible variables for each word were constructed based on research of the literature. Emotional reactivity and the ability to reorganise emotions would have no effect on the participants' descriptions of their perceptions of the phrases 'emotion design', 'emotional design', and 'emotionalise design' and would obviate the effects of gender differences. As a consequence, the gender difference in terms of emotion had little effect on the questionnaire's results.

4.2.2 INSTRUMENTS

A questionnaire was designed (Figure 4.3) to collect quantitative evidence based on numerous sources of information from prior theoretical work. According to Jon and Greene (2003), information was obtained from participants and responders via

Emotion Design, Emotional Design, Emotionalise Design Questionnaire

This questionnaire will assess your knowledge of the following three terms: 'emotion design', 'emotional design', and 'emotionalise design'.

There are no correct or incorrect responses in this level regarding your comprehension of these three phrases. It is entirely subjective as to what you believe you know.

Please kindly √ the applicable boxes.

1. Have you ever heard of the term 'emotion design'?

    • Yes

    • No

2. Have you ever heard of the term 'emotional design'?

    • Yes

    • No

3. Have you ever heard of the term 'emotionalise design'?

    • Yes

    • No

4. The phrase 'emotion design' should be used in reference to the following (you may pick more than one option):

    • Designer-driven (由設計師角度出發)

    • Users/consumers-driven (由使用者/消費者角度出發)

    • Design outcome (設計成果)

    • Design process (設計過程)

FIGURE 4.3   Questionnaire for Empirical Study 1 – quantitative research

- Design consumption (設計成品的消耗過程)

- Decision-making (決策)

- Visceral level (本能維度: 物品的美觀度)

- Behavioural level (行為維度: 使用物品的所帶來的果效)

- Reflective level (反思維度: 物品所引起對自我形象/記憶/個人滿足感)

- Emotional experience (情感經驗)

- Others (please clarify) _____

5. The phrase 'emotional design' should be used in reference to the following (you may pick more than one option):

- Designer-driven (由設計師角度出發)

- Users/consumers-driven (由使用者/消費者角度出發)

- Design outcome (設計成果)

- Design process (設計過程)

- Design consumption (設計成品的消耗過程)

- Decision-making (決策)

- Visceral level (本能維度: 物品的美觀度)

- Behavioural level (行為維度: 使用物品的所帶來的果效)

- Reflective level (反思維度: 物品所引起對自我形象/記憶/個人滿足感)

- Emotional experience (情感經驗)

- Others (please clarify) _____

6. The phrase 'emotionalise design' should be used in reference to the following (you may pick more than one option):

- Designer-driven (由設計師角度出發)

- Users/consumers-driven (由使用者/消費者角度出發)

FIGURE 4.3   Continued

• Design outcome (設計成果)

• Design process (設計過程)

• Design consumption (設計成品的消耗過程)

• Decision-making (決策)

• Visceral level (本能維度: 物品的美觀度)

• Behavioural level (行為維度: 使用物品的所帶來的果效)

• Reflective level (反思維度: 物品所引起對自我形象/記憶/個人滿足感)

• Emotional experience (情感經驗)

• Others (please clarify) _____

7. Have you ever studied any of the following design styles?

• Emotion design

• Emotional design

• Emotionalise design

• I don't know

8. Have you ever designed using any of the following design styles?

• Emotion design

• Emotional design

• Emotionalise design

• I don't know

Gender: F/M

–The end–

FIGURE 4.3   Continued

a series of questions, and statistical analysis of the replies was performed. The case study data offered quantitative descriptions of inexperienced design students' (i.e. future designers') overall conceptions of design and emotion and how inexperienced design students assessed emotion in their design experiences.

A descriptive survey was created according to the aims of the investigation. (According to Lodico, Spaulding, and Voegtle (2010), a descriptive survey provides information and characteristics about a phenomenon by addressing the who, what, where, when, and how questions.) It was a one-time survey in which participants were asked to rate their impressions of the phrases 'emotion design', 'emotional design', and 'emotionalise design' at a certain moment in time. We examined two research questions:

1. How do inexperienced design students interpret the phrases 'emotion design', 'emotional design', and 'emotionalise design'?
2. How can inexperienced design students comprehend the emotional component of design?

The two questions used verbs ('believe' and 'perceive') to elicit inexperienced design students' ideas and feelings about design and emotion. Based on the general research topic stated above, several detailed questions were designed to incorporate the survey's main components:

1. Had the inexperienced design students ever heard of the phrases 'emotion design', 'emotional design', and 'emotionalise design'?
2. Did the inexperienced design students comprehend the phrases 'emotion design', 'emotional design', and 'emotionalise design' as defined in this literature study on design and emotion?
3. Did the inexperienced design students understand the impact of emotion on design (i.e. design outputs, decision-making, and the design process)?

Following a study of the literature on design and emotion, possible factors were found for the phrases 'emotion design', 'emotional design', and 'emotionalise design'. These three phrases correspond to the three primary components of Ho and Siu's 3E paradigm (2009a). According to Desmet and Hekkert (2009), design and emotion may be classified into three broad categories that depend on the designers, design results, and users'/consumers' responsibilities in the design process, design consumption, and emotional experience. Additionally, the notion of 'emotion design' may be defined in terms of the three levels of information processing involved in design consumption (visceral, behavioural, and reflective) (Ho and Siu, 2009a; Norman, 2004). Ho and Siu (2009a) proposed that designers' emotions influenced various decision-making processes during the design process and therefore influenced the development of the design results. Thus, the terms related to the roles of designers, users/consumers, and design outcomes were used as possible variables to ascertain undergraduate design students' perceptions of the terms 'emotion design', 'emotional design', and 'emotionalise design'. The questions that were asked encompassed the concepts 'design output', 'design process', 'design consumption', and so forth. The replies of each responder were analysed to establish the inexperienced undergraduate design students' overall impressions of emotional issues in their design research.

4.2.3 RESULTS

The results indicated that the majority of inexperienced undergraduate design students lacked basic knowledge of the emotional dimensions of design. They were unaware of the connection between emotion and the designers, users/consumers, and design results. Few students were familiar with the terminology associated with the umbrella phrase 'design and emotion'.

Inadequate Understanding of All Three Terms

Only 3.3% of respondents were familiar with all three words. In comparison to emotionalise design, more respondents had heard of emotion design and emotional design, with 40.8% having heard of emotion design and 40% having heard of emotional design. Only 3.3% of respondents were familiar with the term 'emotionalise design'.

Inadequate Understanding of Particular Terms

Over 70% of respondents were unable to link appropriate subjects with specific words. They found few distinctions between the three designations. Only 31.4% of respondents connected the term 'emotion design' to the visceral, behavioural, or reflective levels. Additionally, 37.5% of respondents connected the term 'emotional design' with terms such as 'users/consumers-driven', 'design consumption', and 'emotional experience'. Nevertheless, other respondents believed that emotional design was also linked with the 'design process' and/or 'decision-making'. Additionally, they were unaware of why and how emotion affects decision-making and design processes. Only 21.1% of respondents connected the term 'emotionalise design' with the terms 'designer-driven', 'design process', and 'decision-making'. By and large, respondents believed that the three phrases 'emotion design', 'emotional design', and 'emotionalise design' were more closely associated with users/consumers-driven research than with designer-driven research.

Inadequate Understanding of the Influence of Emotional Concerns on Design

Seventy-four per cent of respondents were unsure if they had ever designed an emotion design/emotional design/emotionalise design or not, whereas 80.2% were unsure whether they had ever designed an emotion design/emotional design/emotionalise design or not.

TABLE 4.1
Research for Empirical Study 1 — Quantitative Research

Question	Response Rate	Analysis
1. Have you ever heard of the term 'emotion design'?	• Yes (49 responses, 40.8%) • No (71 responses, 59.2%)	**Overall Understanding of Design and Emotion** • Most of the undergraduate design students do not have much understanding of the emotional concerns in design study. • They do not understand the relationship between emotion and the role of designers, users/consumers, and design outcomes. • Not many students had heard the terms under the umbrella term 'design and emotion'. **The Students' Understandings of the Three Terms** • 59.2% of students had never heard of 'emotion design'. • Only 31.4% of students were able to associate emotion design with the visceral, behavioural, or reflective levels. • 60% of students had never heard of emotional design. • Only 37.5% of students were able to associate emotional design with users/consumers-driven, design consuming, or emotional experience. • 96.7% of students had never heard of emotionalise design. • Only 21.1% of students were able to associate emotionalise design with designer-driven, design process, or decision-making. • Although there were some students who were able to associate the three main components of the 3E model (i.e. emotion design, emotional design and emotionalise design) with related roles (i.e. designers, users/consumers), it is not yet clear that they understood the definitions of those three main components of the 3E model.
2. Have you ever heard of the term 'emotional design'?	• Yes (48 responses, 40%) • No (72 responses, 60%)	
3. Have you ever heard of the term 'emotionalise design'?	• Yes (4 responses, 3.3%) • No (116 responses, 96.7%)	

(Continued)

TABLE 4.1 (Continued)
Research for Empirical Study 1 — Quantitative Research

Question	Response Rate	Analysis
4. The phrase 'emotion design' should be used in reference to the following (you may pick more than one option):	• Designer-driven (47 responses, 14.5%) • Users/consumers-driven (47 responses, 14.5%) • Design outcome (23 responses, 7.1%) • Design process (18 responses, 5.5%) • Design consuming (6 responses, 1.8%) • Decision-making (7 responses, 2.2%) • Visceral level (37 responses, 11.4%) • Behavioural level (29 responses, 8.9%) • Reflective level (44 responses, 13.5%) • Emotional experience (55 responses, 16.9%) • Others (12 responses, 3.7%) (Content: I don't know, no idea)	
5. The phrase 'emotional design' should be used in reference to the following (you may pick more than one option):	• Designer-driven (25 responses, 8.4%) • Users/consumers-driven (48 responses, 16.2%) • Design outcome (21 responses, 7.1%) • Design process (16 responses, 5.4%) • Design consuming (15 responses, 5.1%) • Decision-making (10 responses, 3.4%) • Visceral level (27 responses, 9.1%) • Behavioural level (29 responses, 9.8%) • Reflective level (37 responses, 12.5%) • Emotional experience (48 responses, 16.2%) • Others (21 responses, 7.1%) (Content: I don't know)	

TABLE 4.1 (Continued)
Research for Empirical Study 1 — Quantitative Research

Question	Response Rate	Analysis
6. The phrase 'emotionalise design' should be used in reference to the following (you may pick more than one option):	• Designer-driven (18 responses, 7.3%) • Users/consumers-driven (23 responses, 9.3%) • Design outcome (18 responses, 7.3%) • Design process (17 responses, 6.9%) • Design consuming (15 responses, 6%) • Decision-making (17 responses, 6.9%) • Visceral level (13 responses, 5.2%) • Behavioural level (19 responses, 7.7%) • Reflective level (35 responses, 14.1%) • Emotional experience (25 responses, 10.1%) • Others (48 responses, 19.4%) (Content: I don't know, redesign some design which is not emotional, don't care, same things)	
7. Have you ever studied the following design style?	• Emotion design (13 responses, 10.6%) • Emotional design (19 responses, 15.4%) • Emotionalise design (0 responses, 0%) • I don't know (91 responses, 74%)	**The Students' Experiences of the Affection of Emotional Concerns in Design** • Many students were not sure if they had experienced the influence of emotional concerns in design: 74% of participants were not sure whether they had ever consumed an emotion design/ emotional design/ emotionalise design or not.
8. Have you ever designed using the following design style?	• Emotion design (7 responses, 5.8%) • Emotional design (15 responses, 12.4%) • Emotionalise design (2 responses, 1.7%) • I don't know (97 responses, 80.2%)	• 80.2% of participants were not sure whether they had ever designed an emotion design/ emotional design/ emotionalise design or not.

(Continued)

TABLE 4.1 (Continued)
Research for Empirical Study 1 — Quantitative Research

Question	Response Rate	Analysis
9. Gender:	• F (68 responses, 56.7%) • M (52 responses, 43.3%)	**The Influence of Gender Difference on the Research Result** • In general studies of emotion, gender differences are one of the factors that may influence the result. • Compared with males, females were more emotionally reactive. They generally performed better on the reorganisations and descriptions of their own emotional changes. • However, the gender differences were not the factor that affected the result as the set of potential variables of each term were developed. The variables provided convincing descriptions for the terms.

4.3 EMPIRICAL STUDY 2: QUALITATIVE RESEARCH

This qualitative research was a focus group discussion with teachers on how to keep children emotionally engaged throughout the design process to work effectively.

4.3.1 PARTICIPANTS

To gain more complete knowledge of the link between inexperienced design students' emotions and the design process to explain the data gathered in the first stage via questionnaires, an in-depth focus group was conducted in this second stage of the study (Figure 4.4). Additionally, this qualitative research examined students' challenges as they managed their design processes to better understand their emotional links with design and how emotion influences their decision-making and design processes and hence design results. Eight individuals were randomly chosen from Empirical Study 1 and invited to the focus group as well. The study's participants were all inexperienced design students enrolled in degree programmes in Hong Kong. A group of eight ensured that adequate input would be supplied; a smaller group might have been unable to provide adequate information. Additionally, if one or two persons dominated the debate or were hesitant to contribute in a smaller group, there would have been less interaction among the participants. Additionally, the larger grouping would improve interaction and group dynamics and result in more accurate feedback. If the focus group discussions contained more than eight members, the researcher would have difficulty managing the dynamics. Similarly, if there were too many participants, it would be more challenging for researchers, since they would have to spend time investigating each individual and asking general questions one by one.

To provide a gender balance, four males and four females were randomly selected and invited from the Empirical Study 1 sample. As with Empirical Study 1, gender

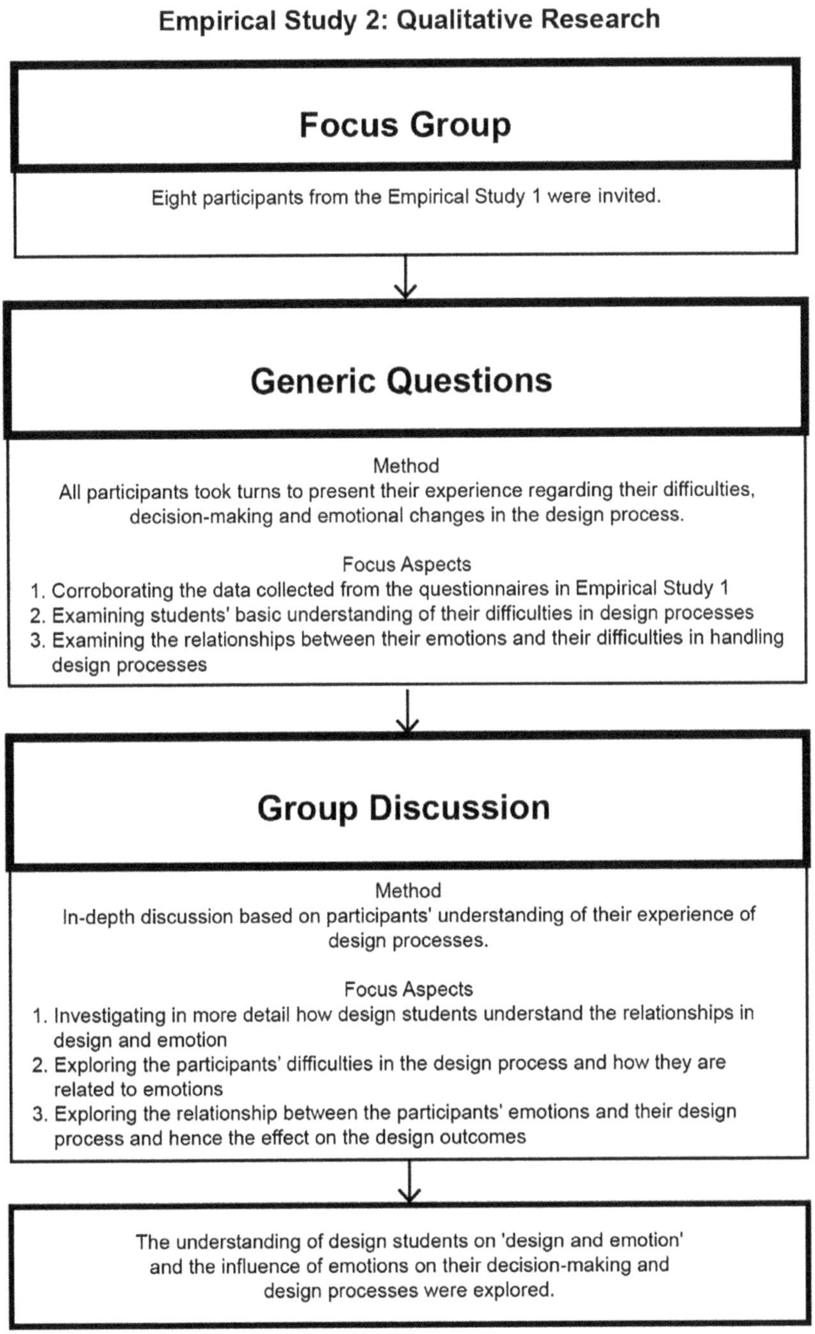

Empirical Study 2: Qualitative Research

Focus Group
Eight participants from the Empirical Study 1 were invited.

↓

Generic Questions
Method All participants took turns to present their experience regarding their difficulties, decision-making and emotional changes in the design process. Focus Aspects 1. Corroborating the data collected from the questionnaires in Empirical Study 1 2. Examining students' basic understanding of their difficulties in design processes 3. Examining the relationships between their emotions and their difficulties in handling design processes

↓

Group Discussion
Method In-depth discussion based on participants' understanding of their experience of design processes. Focus Aspects 1. Investigating in more detail how design students understand the relationships in design and emotion 2. Exploring the participants' difficulties in the design process and how they are related to emotions 3. Exploring the relationship between the participants' emotions and their design process and hence the effect on the design outcomes

↓

The understanding of design students on 'design and emotion' and the influence of emotions on their decision-making and design processes were explored.

FIGURE 4.4   The qualitative research structure for Empirical Study 2

differences are always examined because there may be confounding variables in general studies of emotion; females are more emotionally reactive than males (Bradley et al., 2001; Schimmack, Oishi, and Diener, 2002). However, the results of Empirical Study 1 indicated that there was no difference in the feedback received from the two distinct genders. In this empirical study, a balanced gender ratio was used, and the researcher actively pushed participants to organise and explain their emotional experiences in design throughout the focus group's interactive talks.

Design students were picked over designers to ensure they had at least some experience developing their own design ideas but lacked expertise in managing design projects autonomously. Given their inexperience with design, it was assumed that their responses would be spontaneous and unaffected by other possible influences, such as skill and knowledge transfer through instructors and reference books. Additionally, while the fundamental teachings focused on design thinking and methodology, participants gained an understanding of the design process, the position of designers and design outcomes, and the role of consumers in the context of design research, making them excellent candidates for the study.

To ensure that participants experienced only moderate emotion and did not become overly enthusiastic or anxious, which could have impacted the outcome, the researcher explained the study's objective and that participants' confidentiality would be honoured before the start of the study. Additionally, he emphasised that no correlation existed between their conversations or critiques and their academic performance in the design study.

4.3.2   PROCEDURES

The researcher and all participants were seated in a circle to allow the researcher to view each member of the focus group. The researcher and all participants had face-to-face talks to foster group engagement. At the start of the interview, the researcher explained the study's purpose and procedures to familiarise participants with the subject and prepare them to share their experiences during the study's two sections: generic questions (Figure 4.5) and group discussion (Figure 4.6). The entire research session lasted about one-and-a-half hours, since participants were required to reflect on the questions without being distracted by lengthy discussion. The interviews were done in Cantonese to enable students to respond more freely and express themselves fully.

Generic Questions

To corroborate the data collected via questionnaires in the first empirical study and to examine students' fundamental understanding of their difficulties with design processes as well as the connections between their emotions and their difficulties with design processes, the researcher asked generic questions during the first segment of the focus group. The questions centred on the challenges, decision-making, and emotional shifts experienced by students during the design process. Each participant presented their experience to the researcher in turn. After responding to a series of basic questions about their own experiences with design and emotion, it was

Part 1. Generic Questions

1. Have you ever encountered challenges during the design process, such as those shown in the following

    situations? (Students may select many scenarios)

    • lack of information

    • lack of communication between the team members

    • poor emotion management

    • poor time management

    • lack of materials

    • poor risk management

    • other (please specify)

    在以往的設計過程中，你有否遇過以下困難？（可選多於一個）

    • 缺乏資訊

    • 與組員缺乏溝通

    • 情緒管理技巧不足

    • 時間管理技巧不足

    • 缺乏材料

    • 危機管理技巧不足

    • 其他（請說明）

2. Would you explain your own encounters with the challenges you selected in question 1?

    能否形容你如何經歷在問題一中所指出的困難嗎？

3. What are your thoughts about emotion?

    你認為甚麼是情感？

FIGURE 4.5   Questions discussed in the focus group (Part 1: Generic questions)

4. With reference to the challenges you encountered during the design process (in question 1), did they lead you to experience any emotional changes? If you answered affirmatively, could you match the emotional shifts to the following eight categories of emotion?

- aroused,   astonished,   stimulated,   surprised,   active

- enthusiastic,   elated,   excited,   euphoric,   lively,   peppy

- happy,   delighted,   glad,   cheerful,   warm-hearted,   pleased

- relaxed,   content,   at rest,   calm,   serene,   at ease

- quiet,   tranquil,   still,   idle,   passive

- dull,   tired,   drowsy,   sluggish,   bored,   droopy

- unhappy,   miserable,   sad,   grouchy,   gloomy,   blue

- distressed,   annoyed,   fearful,   nervous,   jittery,   anxious

對於你在問題一中所指出的困難，他們有否令你有任何情感改變？ 如有，請從以下八組情感種類分辨出它來。

- 激動，   驚訝，   激勵，   意想不到，   活躍

- 熱情，   興高采烈，   興奮，   心滿意足，   活潑，   精神充沛

- 高興，   快樂，   快活，   愉快，   熱心，   滿意

- 放鬆，   滿意，   靜止，   冷靜，   寧靜，   自在

- 安靜，   平靜，   靜默，   無聊，   不活躍

- 晦暗，   疲倦，   困倦，   懶散，   乏味，   疲乏

- 不高興，   痛苦，   悲哀，   抱怨，   憂鬱，   沮喪

- 惱怒，   氣惱，   害怕，   緊張，   不安，   焦慮

5. Are you aware that your emotional state may have an effect on the subsequent stage of the design process?

你知道你的情感改變會影響該設計過程的下一個步驟嗎？

FIGURE 4.5   Continued

6. Are you familiar with the link between decision-making and design? Are you aware that the design process entails a great deal of decision-making?

你知道作決定和設計過程的關係嗎? 你知道在一個設計過程中, 包括許多不同的決定嗎?

7. Have you ever heard that your emotions may influence your design decisions?

你有否聽過你的情感會影響你在設計過程中所作的決定?

8. Have you ever encountered challenges throughout the design process as a result of variables outside your control/expectations (e.g. societal or technical changes)?

你有否遇過任何困難是你所不能控制/在你預料之外?

Are you aware of the connection between emotion and design?

If you answered affirmatively, what is/are the reason(s) for your understanding of the link between emotion and design?

If not, for what reason(s) do you believe you lack a strong grasp of the link between emotion and design?

你知道情感和設計之間有甚麼關係嗎?

如知道, 你認為是甚麼原因令你對情感和設計兩者之間的關係有所了解?

如不知道,你認為是甚麼原因令你對情感和設計兩者之間的關係了解不多?

9. Do you believe your design study includes sufficient consideration of emotion and design?

你認為在設計教學中,對情感和設計兩者的關係的討論足夠嗎?

FIGURE 4.5 Continued

Part 2. In-depth Discussion

1. Have you ever encountered the power of emotion as it influences your decision-making? How did
    it affect your mental state in the subsequent stage?

    在以往的設計過程中,你曾否在情感影響下作決定? 這決定如何影響該設計過程的下一個步
    驟?

2. Have you ever been subjected to the sway of emotion at any point in the creative process? How did
    it affect the outcome/result of your design?

    在以往的設計過程中,你曾否受情感影響? 這決定如何影響該設計成品/結果?

3. Are you aware of the importance of emotion in the design process? If yes, how did you learn that?

    你知道情感在設計過程中擔當何種角色嗎? 如知道,請解釋你如何得知。

4. What link do you believe exists between design and emotion?

    你認為設計和情感有何關係?

5. Do you believe that including emotion in your design research will be beneficial?

    你認為情感會對你的設計學習有幫助嗎?

FIGURE 4.6   Discussions in the focus group (Part 2: In-depth discussion)

anticipated that participants would have a better understanding of the subject and would debate more in-depth topics in the subsequent group session.

Group Discussion

To explore the effect of emotion on inexperienced design students' decision-making, design processes, and hence design results, an in-depth conversation (driven by a series of questions) was held. To provide participants sufficient time to recollect their experiences, the researcher invited them to consider the questions for a few minutes before initiating a fluent conversation. The participants were encouraged to engage in participatory presentations of their expertise and experience.

4.3.3 INSTRUMENTS

A semi-structured interview (i.e. qualitative research) was utilised to clarify the data gathered and to gain a more thorough knowledge of the results of Empirical Study 1. Semi-structured interviews, according to Fraenkel and Wallen (2008), are more formal in style and include a series of questions meant to elicit specific replies from respondents. All of the data gathered may be compared and contrasted afterwards. Given the study's objective, standardised open-ended interviews were used in conjunction with certain closed-response questions. To prepare for the standardised open-ended interviews, predetermined questions and response categories were used (Wallen and Fraenkel, 1991). The same fundamental questions were asked of all participants in the same sequence. The majority of questions were phrased in an open-ended manner. Due to the fact that all participants answered identical questions, the replies were more comparable; data on the interview questions were collected for each participant. When several interviewers were utilised, interview effects and bias were decreased. During the assessment stages, the equipment was examined; data organisation and analysis were facilitated. Simultaneously, to generate replies that could be readily compared and aggregated, certain restricted-response questions were developed as a supplemental tool.

Research Questions

The following three research questions served as the basis for both the general questions and the group discussion:

Q1. What are your design challenges?
Q2. What is your interpretation of the link between the emotions you have had and the problems you have encountered during the design process?
Q3. What is your experience and knowledge of emotion's effect on your decision-making, design process, and hence on the design outcomes?

On the basis of the three general research topics above, more detailed questions were generated to form the survey's main parts.

The first research question, 'What are your design challenges?' evolved into the following two:

- Have you ever encountered challenges during the design process, such as those illustrated in the following scenarios?
- Are you able to relate your personal encounters with such difficulties?

The second research question, 'What is your interpretation of the link between the emotions you have had and the problems you have encountered during the design process?' resulted in the following five questions:

- What are your thoughts about emotion?
- Are you aware of the connection between emotion and design? What is/are the reason(s) for your lack of knowledge about the link between emotion and design?
- Is it the result of insufficient discussion throughout your design study?
- What link do you believe exists between design and emotion?
- Do you believe that including emotion in your design research will be beneficial?

The third research question, 'What is your experience and knowledge of emotion's effect on your decision-making, design process, and hence on the design outcomes?' resulted in the following seven questions:

- With regard to the challenges that you encountered during the design process; did they cause you to undergo any emotional changes?
- Are you aware that your emotional state may have an effect on the subsequent stage of the design process?
- Are you familiar with the link between decision-making and design? Are you aware that the design process entails a great deal of decision-making?
- Have you ever heard that your emotions may influence your design decisions?
- Have you ever encountered the power of emotion to influence your decision-making? Did it affect your emotional state in the subsequent stage?
- Have you ever been subjected to the sway of emotion at any point in the creative process? How did it affect the outcome/result of your design?
- Are you aware of the importance of emotion in the design process? Why are you aware of this?

The critical incident method previously introduced by Flanagan (1954) was used to answer the first ('What are your difficulties in the design process?') and third ('What is your experience and understanding of the influence of emotion on your decision-making, design process, and thus the design outcomes?') research questions. The participants were asked to identify and characterise design process obstacles as well as the impact of emotion on their decision-making and design processes. This approach is considered superior to standard surveys and observations because it produces detailed, rich tales that are typically more informative and applicable to design practice.

To investigate the relationship between participants' emotions and their design processes, they were asked to recall their encountered difficulties and then describe and differentiate the corresponding emotion in response to the second research question (i.e. 'How do you understand the relationship between your emotions and your design difficulties?'). It was anticipated that participants would demonstrate their understanding of the function of emotions by categorising emotions according to their 'underlying dimensions'. Desmet (2002) used a similar technique to categorise emotions in his product emotion study in which emotions are classified according to their manifestations, prior evaluations, and underlying dimensions. Because emotions are linked and some are more similar than others (e.g. rage vs annoyance), they are best defined and distinguished using underlying qualities. Generally, the aspects are referred to as 'pleasantness' and 'activity'. The 'pleasantness' dimension spans the spectrum from unpleasant (e.g. sad) to pleasant (e.g. happy). The dimension 'activation' is described as physiological arousal (Clore, 1994), and it encompasses states ranging from calm (e.g. content) to enthusiastic (e.g. euphoric). Emotions may be divided into eight categories (neutral excited, pleasant excited, pleasant average, pleasant calm, neutral calm, unpleasant calm, unpleasant average, unpleasant excited) using these 'pleasantness' and 'activation' dimensions (Russell, 1980). In this study, participants were asked to explain their emotional shifts in terms of linked emotions and how these emotions influenced their decision-making and creative processes. As a result, emotion was classified according to its fundamental characteristics.

4.3.4 RESULTS

The authors examined the links between new-generation designers' emotions and decision-making. Emotions are constantly used in decision-making, and the design process involves a great deal of decision-making. However, no design student understands the connection between emotion and the design process. They are unsure of the right method to include emotion in their creative processes. Some design students have developed self-initiated techniques for resolving emotional issues, while others continue to struggle with the design process. As a result, there is a need to examine the relationship between design and emotion in general design courses.

Emotions Caused by Difficulties in the Design Process

Seventeen cases illustrating the primary difficulties in the design process of designers at the entry level (e.g. insufficient time management, insufficient information, insufficient risk management, insufficient communication with team members/tutors, and insufficient emotional management) were identified using the critical-incidents technique (Guba and Lincoln, 1981; Miles and Huberman, 1984). Following that, participants identified the mood elicited by those obstacles. The majority of individuals characterised their emotions as 'unpleasant eager', 'unpleasant calm', or 'unpleasant average'. Distressed, irritated, scared, tense, restless, and worried were all mentioned 14 times in the 'unpleasant excited' category. Nine other emotions were noted in the 'unpleasant calm' category, including dull, weary, drowsy, sluggish, bored, and drooping. The 'unpleasant average' group had seven emotions: unhappy, wretched, sad, grumpy, gloomy, and blue. The feeling of 'neutral excited' was stated six times. It comprised aroused, amazed, stimulated, surprised, active, and intense. The feeling of 'neutral calm' was stated four times. It comprised quiet, peaceful, motionless, inactive, idle, and passive. The adjectives enthusiastic, exhilarated, eager, euphoric, vivacious, and peppy were all used three times in the 'pleasant excited' category. No participants mentioned the category of 'pleasant average' feelings (glad, thrilled, etc.) or the category of 'pleasant calm' emotions (relaxed, content, etc.). For instance, one of the participants reported an encounter that was handled ineffectively: 'In the group-design project, a teammate arrived late for a meeting, logged on to Facebook, and talked during the meeting'.

This incident describes a design process difficulty that the participant coded as 'other: lack of coordination with team members' but which should have been coded as 'lack of communication with team members' because the most significant aspect of the difficulty was a failure to communicate the agreed schedule among team members during the design process. The participant noted that he was experiencing 'blue' feelings as a result of this problem.

Confused Understanding of Emotion

Participants' sense of emotion was amorphous, and they lacked a precise definition; their interpretations included feelings, mood, and attachment. One participant said that emotion affected human behaviour; another stated that emotion is a form of expression for individuals. According to another participant, it is a type of response to a stimulus. Simultaneously, one participant said that emotion characterised a person's feelings; another claimed that emotion is a 'component' (i.e. technique) of how individuals develop their connections.

Emotional Work as a Catalyst to a Design Process

Five participants had been told that their emotions would influence their design decisions. All participants recognised that their emotional states would have an effect on the subsequent stage of the design process, but they were unaware of how emotion influenced it. They provided some examples of how emotion plays a role in the design process. Certain participants indicated that distinct emotions might serve distinct purposes at various stages of the design process. Additionally, emotion was regarded as a trigger that might hinder or enhance designers' decision-making abilities. Designers' decision-making talents would impede or improve the impact of decision-making on the design process, since decision-making constructed each step of the design process.

Various Functions of Emotions in the Design Process

Several participants stated that emotion may play a role in the design process as well. Several of them believed that emotions would serve a distinct purpose at each level of the design process, impairing or enhancing designers' decision-making abilities. While all participants encountered the effects of emotion on decision-making, their experiences were seldom linked with the idea of design and emotion. Several of them discovered that emotion can influence decision-making in the short term. According to some, it had a substantial impact on decision-making, which determined the effectiveness of design outputs, and pleasant emotions should benefit designers in making effective judgements. According to some others, emotions have a detrimental influence on decision-making because they are too subjective. In any case, all these students claimed that emotion plays a big role in the designer's distinctive style, therefore amplifying the influence of design decision-making. Participants stated that emotion might be included in the design process either as a trigger or a co-creator.

Emotion Affects Decision-making and the Design Process

All participants recognised the link between decision-making and the design process; they recognised that every design process involves significant decision-making. The majority of participants were aware that emotion influenced their daily decisions but were unaware that it also applied to the design process. Six of the participants encountered challenges throughout the design process as a result of events outside their control or expectations. The majority of these things were brought about by technical advancements. One participant said that emotion regulation is one of the issues that will complicate his design process and is uncontrollable.

Inadequate Discussions on Emotion and Design in Design Studies

All participants believed that there is insufficient discussion in design research regarding emotion and design. One student noticed that while certain areas of their design study impacted on 'emotion and design', there were no lectures that concentrated directly on it. One student stated that while certain books and seminars discussed emotion in conjunction with other topics, design lessons made no such reference to emotion.

Inadequate Understanding of How Emotion Affects the Design Process

All participants encountered the effects of emotion during the design process, but they lacked an understanding of how emotion affects the outcome/result of their design. The group project immediately recognised the importance of emotion in the design process. The way team members communicate will have an effect on their emotions during the design process. The influence of emotion on design processes is a feedback loop, which was especially evident throughout the group project design process.

Positive and Negative Influences on the Design Process

Participants suggested that emotions would have both positive and negative influences on design processes. In light of their experience of design practices, they suggested that positive emotions would be helpful to balance junior designers' lack of experience. The participants stated that designers have to manage their emotions in the design process, and they were interested to learn more about emotion management to solve the difficulties in design processes and to be more creative. Some participants suggested that emotion would also influence the message of design outcomes.

Emotion as a Possible Topic in the Reform of Design Studies

The participants were generally unaware of the connections between emotion and design, while the majority believed that such relationships exist. They believed that emotion can aid in the development of designers' styles. They recognised that emotion has an effect on their daily decisions, but they had never connected it to their creative process. They stated that they would use strategies such as listening to songs, reading books, and watching films to help them regulate their personal emotions to prepare for everyday decision-making. As a result, all participants believed that including emotion in their design research would be beneficial. Some individuals attempted to moderate their emotions but were unsuccessful in some instances. They were unable to articulate the causes of their emotion management failures. Several participants recognised that emotion would be evident in students' thoughts and would influence their decision-making in a similar design situation in the future. They agreed that emotions would help them to diversify their problem-solving thought pathways.

4.4   ROLE OF RESEARCHER

The researcher questioned the participants and requested them to respond one by one to the generic questions. Participants were asked to express their own thoughts in response to one another's responses. The researcher then facilitated an in-depth discussion with the participants using a series of questions. The researcher used an audio recorder and photographed the chats and also analysed the data for the entire study. The advantage of documenting facial expressions, body gestures, and dialogue is that photographs and audio convey detailed information about actual situations as perceived by individuals (Creswell, 2008).

4.5   BENEFITS AND LIMITATIONS

4.5.1   BENEFITS

Course Reorganisation in the Light of Emotions

Emotions induced by design issues were classified as 'neutral excited', 'pleasant excited', 'neutral calm', 'unpleasant calm', 'unpleasant average', and 'unpleasant excited'. As a result, unpleasant feelings were frequently encountered during the design process. In contrast, several participants believed that pleasant emotions might be beneficial in compensating for junior designers' lack of design expertise. Participants shared how they change unpleasant feelings into positive ones throughout the group discussion. Due to the fact that their sharing was limited to personal experiences and the sample size was small, the results cannot be used to produce broad suggestions for increasing emotion in the design process. Additional research is needed to determine how designers at the entry level might change negative feelings into positive ones.

The Relationships Between Emotions, Decision-making, and the Design Process

The authors examined the relationship between emotions and decision-making in the current generation of designers. Students had always used emotions in their decision-making during design processes, but they were unaware of their significance in design. They had encountered the interplay between emotion and design and had included concepts of design and emotion in their design processes, but they lacked a thorough understanding of the subject.

The First Study to Explore the Perceptions of Designers

To the authors' understanding, this has been the first study to examine designers' perceptions of the phrases 'emotion design', 'emotional design', and 'emotionalise design' as well as their understanding of the role of emotion in design outcomes, decision-making, and design processes. This exploratory project examined the importance of educating the next generation of designers about the significance of emotion in design.

4.5.2 LIMITATIONS

Limitations of the Study Sample

Due to resource constraints, all respondents were Hong Kong students. Salili and Hoosain (2001) assert that Hong Kong pupils are less self-sufficient and rely substantially on their teachers. Additionally, teachers prepare the majority of their educational experiences. They are not like Western pupils, who are encouraged to develop their own learning processes and outcomes. Students in Hong Kong and Western countries have significantly distinct learning styles; Hong Kong students place a higher premium on skill development. They engage in less self-discovery and reflection on their experiences. The knowledge of emotion in design studies may vary globally for inexperienced design students/designers at the entry level. In comparison to Western pupils, Hong Kong students clearly lack an appreciation for the role of emotional concerns in design. However, as design education continues to globalise, Hong Kong design education is incorporating more Westernised educational approaches (Sutherland, 2002). Due to the growing popularity of cross-cultural studies in Hong Kong design education, multicultural design teaching and learning approaches are being implemented (Biggs and Watkins, 2001). The divide between Hong Kong and Europe in terms of design education is decreasing. Additional research is required to determine the difference in knowledge between Hong Kong and Western students. Another constraint is the tiny sample size. Due to resource constraints, only eight people could be enlisted.

Limitations of the Questions Discussed

Although the questionnaire and group discussion questions were kept as easy as possible, participants were still required to interpret the questions based on previous experiences and knowledge of design. Other students' interpretations of the design terms 'design process', 'decision-making', and so on may differ. Throughout this study, participants' interpretations of those phrases served as the foundation for defining their design challenges and their experiences with how emotions affect their design processes. It became apparent that their disparate conceptualisations of those design terms may have generated disparate descriptions of their knowledge of design and emotion.

Limitations of the Research Method

Due to resource constraints, an action study on how designers at the entry level deal with emotions during their decision-making and design processes was not possible.

Action research, in which people are invited to collaborate on a design project, may assist researchers in identifying the obstacles encountered by junior designers and their emotional shifts during the design process. While action research is a successful approach for observing participants' performance instead of criticising and making comparisons of their knowledge of emotion, decision-making, and the design process, it cannot provide information about how the student recognises the role of emotion and its significance in the design process.

4.6 SUMMARY

This is a survey of the current generation of designers' broad grasp of the importance of emotions in decision-making and the design process as well as its effect on design outcomes. This is a sampling study; participants are studying a variety of disciplines within graphic and multi-media design. While some are studying animation and others graphic design, the results indicate that they are all confronted with similar difficulties surrounding emotion during the design process. Similarly, the same observations apply to other facets of graphic design. In the 1990s, design professions began to investigate the relationship between design and emotion; an equivalent terminology with clear and specific logic and definition was formed. However, there is a dearth of discussion in design studies on the topic of 'design and emotion'. This study demonstrates that the majority of inexperienced design students are unaware of the connection between design and emotion. Additionally, they lack an understanding of why and how emotion affects decision-making and design methodology. The study's results indicate that there is a dearth of conversation about design and emotion in design education and that inexperienced design students conduct an insufficient examination of the relationship between emotion and design, which would help them to manage their creative processes. To inspire new generations of designers to appreciate the value of emotion and its role in design, design and emotion are critical in the reform of design education.

5 Discussion and Limitations

5.1 INTRODUCTION

Scholars have concentrated on the design process throughout the history of design practice, examining diverse elements such as the management of the entire design process and the numerous design approaches (Chen and Chen, 2004; Cross and Sivaloganathan, 2004; Gayretli and Abdalla, 1999; Noble and Bestley, 2005; Peto, 1999). However, few studies have examined why inexperienced design students/designers at the entry level are unable to successfully manage their design processes. While tutors offer inexperienced design students/designers at the entry level with valuable instructions for managing their design processes stage by stage, not all tutors are capable of properly implementing the design process. Several of the inexperienced design students/designers will leave the design project if they become disoriented by the directions necessary to accomplish the projects' objectives or by the difficult design-development process. What reasons contribute to their inability to complete their design tasks on time?

The aim of this study is to generate new knowledge about emotion in the design process. To achieve this aim, the following research objectives were identified and explored in different stages of this study:

1. To highlight the importance of the role of emotion in decision-making and the design process.
2. To investigate novice design students' understanding of design and emotion and how this understanding affects their decision-making in the design process.
3. To understand how emotion affects the design process, including the roles of designer, design outcome, and user.
4. To analyse possible applications of the findings with regard to design and emotion in a new curriculum for design education.

The present investigation commenced with a review of the scholarly literature pertaining to the correlation between affect and the process of design. The literature has observed that the investigation into the correlation between design and emotion has yielded several comparable terminologies and constructs, including but not limited to

DOI: 10.1201/9781003388920-5

'emotionalised design', 'emotional design', and 'emotion design'. Nevertheless, it has been determined that the correlation between design and emotion remains ambiguous within these parameters. The analysis of the implications of different terms in the context of emotion and design was deemed necessary. The study revealed a strong interconnection between the concepts of design, consumption, and reflection, which are anchored on three primary functions, namely designers, design outcomes, and users/consumers. The three primary functions entail a substantial degree of engagement with affective states. Designers incorporate their emotions into the design process when creating design outcomes as observed in their interaction with the design.

The theoretical definitions and concepts of the E-Wheel and 3E model were formulated to facilitate the exploration of the correlation between design and emotion. The E-Wheel model was proposed to elucidate the impact of external factors on the emotional states of designers and the subsequent influence of these emotions on the various decision-making processes associated with the internal factors of the design process (Ho, 2010). The study yielded evidence indicating that the emotions of designers play a crucial and primary role in the design process, influencing their reactions to internal factors throughout the said process. Furthermore, the design results that embody the emotions of designers have been labelled as 'emotionalised design'. The resultant design may potentially stimulate the user to establish connections and to recollect or envision associated occurrences or pertinent experiences in order to evoke affective reactions or induce alterations in his or her emotional state. The phenomenon is commonly recognised as 'emotional design' within academic discourse. The term 'emotion design' was coined to describe the interaction that occurs between designers and users through the outcomes of design. The 3E model was established on the basis of the interrelationships between the novel concepts of emotionalised design, emotional design, and emotion design. The E-Wheel model and 3E model have furnished theoretical definitions and concepts that facilitate the exploration of the connections between design and emotion.

The literature review examined the function of emotion in decision-making and the comprehensive design process. A significant number of inexperienced design students encounter challenges pertaining to the inherent factors within their design procedures. Despite adhering to the rational and cognitive learning methodologies espoused by their design instructors, these students encounter challenges when it comes to making decisions regarding intricate aspects of design. The findings of the literature review indicate that the incorporation of emotion can potentially augment the decision-making capabilities of designers and optimise their design processes. Qualitative and quantitative data on these topics were collected in Empirical Studies 1 and 2; participant feedback and sharing provided a variety of ideas for the study.

The first empirical study employed questionnaires as a means of gathering qualitative data on the perception and comprehension of design and emotion among novice design students. The present investigation delved into the manner in which inexperienced design students undergo the process of design and emotion in both their design creation and design consumption. Based on the results of the questionnaires, it was found that a mere 40% of the participants demonstrated familiarity with the fundamental concepts encompassed by design and emotion, including but not limited to 'emotionalised design', 'emotion design', and 'emotional design'. The study revealed that a significant proportion of the participants (40.8% to be specific) were familiar with the term 'emotion design', while 40% of the participants had knowledge of 'emotional design'. Additionally, a minority (just 3.3%) of the participants were aware of 'emotionalised design'. The findings of the initial investigation suggest that inexperienced design students possess an insufficient comprehension of the emotional considerations within the realm of design studies. Furthermore, a significant proportion of the respondents who were familiar with the aforementioned terminologies exhibited an inability to appropriately link corresponding concepts with the said terms. The individuals expressed uncertainty regarding their potential exposure to emotional considerations in the realm of design; specifically, this was with regard to consumption or creation. In general, the results suggest that inexperienced design students possess a limited comprehension of the concepts of design and emotion.

Empirical Study 2 was conducted as a follow-up investigation to corroborate and further explore the findings of Empirical Study 1 through the use of an in-depth focus group. The study investigated the participants' comprehension and familiarity with design and emotion in the design process as well as the impact of emotions on the decision-making and design processes of novice design students. The focus group was utilised as the primary method of data collection. The examination of the two empirical investigations yielded the findings discussed below.

5.2　DISCUSSION

5.2.1　Understanding Design and Emotion

Few Participants Had Heard and Understood the Particular Terms in Design and Emotion

Although several theories and models on design and emotion have been created in the last decade, the current generation of designers has insufficient knowledge of the link between design and emotion. Since the 1990s, psychological research and other fields such as marketing and communication have motivated design experts to investigate the link underlying design and emotion. Overbeeke and Hekkert (1999) developed the term 'design and emotion' and sought to build tools and methodologies for designers to utilise to create an emotionally worthwhile product–user connection. As additional studies were initiated, a network for discussing design and emotion among design experts became necessary. Numerous studies, models, and theories, such as emotional design, emotionalise design, and others, have been developed and accepted to investigate the links involving design and emotion and the reactions to them as well as to explain how emotion may be successfully employed in design. Following a classification of ideas based on research, this study focuses on three primary roles that influence or are influenced by emotion: designers, design output, and users/consumers throughout the design cycle of the design process, consumption, and reflection (Ho and Siu, 2009b). For example, 'emotionalise design' refers to the method through which designers incorporate their personal feelings in the design process (Ho and Siu, 2009b). 'Emotional design' is defined as 'design that elicits an emotional response from users/consumers' (Norman, 2004). 'Emotion design' is a term that refers to design that incorporates emotional considerations into the interactions between designers and users (Ho and Siu, 2009a; Norman, 2004). Simultaneously, as a result of technological advancements and societal and economic shifts, the designer's job has evolved (Ho, 2010), with designers needing to be motivated and able to manage their own design operations. If designers gained a better understanding of the terminology used in design and emotion and the impact of emotion on design, they would be more equipped to innovate (Desmet and Hekkert, 2009; Ho and Siu, 2009b). Designers with strong emotions are better managers than others in the design process because they can think in divergent directions during the design process (Vosburg, 1998) and can choose the most effective strategies to solve problems, thereby improving the quality of their ideas; designers with strong emotions are superior managers during the design process (Aken, 2005).

However, according to the results of the questionnaire in Empirical Study 1 and the focus group in Empirical Study 2, inexperienced design students/designers at the entry level were confused with the notion of emotion. In Empirical Study 1, few participants had heard of the phrase 'design and emotion'. Few of them were familiar with all three phrases, namely 'emotion design', 'emotional design', and 'emotionalise design'. The majority of pupils were unable to link appropriate subjects with specific words. In Empirical Study 2, the majority of inexperienced design students/designers at the entry level demonstrated a lack of awareness regarding the emotional aspects of design. They were unaware of the connections between emotion and the designers, users/consumers, and design results. In order to investigate the comprehension of the concepts of design and emotion among the participants, it was imperative to ascertain their familiarity with topics that are associated with design and emotion, including emotionalised design, emotional design, and emotion design. In the first empirical study, it was found that a mere 40% of the participants were familiar with at least one of the three terminologies encompassed by design and emotion. In the first empirical study, a minority of participants (less than 4% to be specific) demonstrated familiarity with the three terms under investigation, namely 'emotionalised design', 'emotional design', and 'emotion design'. Regarding the terminology, it was found that 40.8% of the participants were familiar with the concept of emotion design, while 40% were familiar with emotional design. Additionally, a minority of 3.3% had knowledge of emotionalised design. In comparison to the concepts of emotion design and emotional design, the terminology of emotionalised design appeared to be the least familiar. The unexpected research finding pertains to the varying levels of awareness among individuals with regard to the three aforementioned terms. Participants who lacked familiarity with any of the three terms were instructed to proceed to the final section of the questionnaire, whereas those who possessed knowledge of any of the terms were requested to respond to the remaining inquiries in the first empirical study. The subsequent inquiries were formulated to examine the participants' comprehension of the definitions of the three aforementioned concepts. Despite the participants' reported familiarity with the terminology, their comprehension of the terms could not be assumed.

Despite the fact that the participants reported familiarity with the aforementioned terms, it cannot be inferred that they possessed a comprehensive understanding of their respective meanings. Based on the findings of the follow-up inquiries in the first empirical investigation, a majority of the respondents demonstrated an inability to accurately identify the respective definitions of the terms 'emotionalised design', 'emotional design', and 'emotion design', thereby resulting in a conflation of their meanings. The findings indicate that the respondents lacked comprehension regarding the interplay between emotions and the functions of designers, users/consumers, and design outputs. While certain participants were successful in associating the designations of the 3E model – namely, emotionalised design, emotional design, and emotion design – with their corresponding functions – namely, designers, users/consumers, and design outcomes – they were unable to accurately match the terms with their respective definitions as outlined in the literature review. In general, it can be observed that the design students who were new to the field exhibited a deficient comprehension of the specific vocabulary associated with design and emotion, such as 'emotionalised design', 'emotional design', and 'emotion design'. Their acknowledgement of the correlation between emotion and design studies is currently limited.

As per the feedback obtained from the focus group in Empirical Study 2, it can be observed that the participants faced difficulty in comprehending the emotional impact or relevance of design outcomes on the user. Some participants expressed uncertainty regarding the methods employed by designers to incorporate diverse emotional considerations into the design process, which ultimately lead to the attainment of their final design objectives. Furthermore, a limited number of participants observed the correlation between designers and users in relation to design outcomes and the potential for these three components (i.e. designers, users, and design outcomes) to form robust connections within the context of design and emotion research. The findings of Empirical Study 2 indicate that a significant proportion of the participants lacked comprehension regarding the emotional aspects of the design study. Additionally, they exhibited a limited understanding of the interplay between emotions and the responsibilities of designers, users/consumers, and design outcomes. While one of the participants acknowledged a connection between their design study and the field of design and emotion, no formal classes or workshops pertaining to the subject of emotion in design were mentioned during the discussion. In general, the results suggest that the notion of design and emotion represents a novel concept for inexperienced design students. The participants were unable to discern whether they had encountered the impact of emotional considerations in design or not. Furthermore, the study reveals that the participants lacked awareness of the impact of user-centric design, which involves both designers and users in the design process. Additionally, only a minority of participants encountered the role of emotions in shaping their design experiences. The study's participants did not observe the emotional fluctuations in users' feedback or the implicit expressions of users.

5.2.2 FAMILIARITY WITH THE CONCEPT OF EMOTION

To understand how some inexperienced design students lacked an appropriate grasp of design and emotion, we must first understand their level of acquaintance with emotion. According to participant comments in Empirical Study 2, younger design groups lacked a clear understanding of emotion. They defined emotion as a type of sensation/mood. Additionally, they lacked an understanding of the distinction between feeling and affection. One participant noted that emotion has an effect on human behaviour; another noticed that it is a form of expression for individuals. According to another participant, it is a type of response to a stimulus. Additionally, one participant said that emotion characterised a person's feelings; another claimed that emotion is a technique that helps individuals create relationships with others. Indeed, theories about emotion have existed for a long time; according to Scherer (1984), emotion is best understood as a series of interrelated, synchronised changes in the state of the majority of the five organic subsystems: information processing, central nervous system, neuro-endocrine system, autonomic nervous system, and executive central nervous system. To distinguish emotion from other similar terms such as 'affective states' (i.e. what participants referred to as 'affections'), Carlson (1997) defined emotion as brief, sharp waves of emotional changes that occur without conscious effort or reflection and are typically accompanied by increasing physiological changes, such as an increased heart rate. In contrast, an affective state is described as a less strong emotional reaction that is more persistent. The participants had difficulty recognising the nature of emotion and providing an appropriate definition.

5.2.3   Familiarity with Emotional Influence on Everyday Decision-making

Although the majority of design students/designers at the basic level are unfamiliar with the notions of emotion, they are keen to explore the impact of emotion on human thought and behaviour. According to participant comments in Empirical Study 2, 5% of participants had heard that emotion will influence their daily decision-making. One participant described how she uses activities such as listening to music to 'brace up' her own mood for each school day. Another participant discussed how emotion affected his everyday decision-making; he wanted to regulate his emotions via reading books or going to the cinema. According to the participants, inexperienced design students realise that emotion enriches daily life and that increased exposure to emotion influences their decision-making capacity. As a result, people have developed strategies for preparing their emotions to make decisions in their daily lives. Cacioppo et al. (2001) share similar views to the participants about how emotion affects daily life: 'Emotions guide, enrich, and ennoble life; they give significance to everyday existence, and they determine the value placed on life and property' (p. 173). Students, on the other hand, are unaware that emotion may impact their design process and even their design studies.

5.2.4   Emotion Caused by Difficulties in the Design Process

The participants were inquisitive about how their emotions affected their capacity to make decisions during the debate on the link between emotion and decision-making. They stated that while they were suffering negative emotions, they would make unsuccessful decisions; in other words, positive emotions should aid in making effective judgements. In certain situations, poor decision-making also resulted in emotional alterations. Ineffective decision-making during the design process resulted in design problems. The primary challenges encountered by designers at the entry level during the design process (e.g. insufficient time management, a lack of knowledge, inadequate risk management, lack of communication with team members/tutors, and poor emotion management) were highlighted in their descriptions. Following that, the participants recognised the feelings elicited by those challenges. The majority of participants defined the feelings evoked by the difficulties as 'unpleasant aroused', such as upset, annoyed, scared, and tense; 'unpleasant calm', such as dull, sleepy, and drowsy; and 'unpleasant average', such as unhappy and wretched. One participant described experiencing a 'blue' mood as a result of team members failing to communicate on a predetermined timetable throughout the design process. While some participants reported that emotion had a short-term influence on their decision-making, it can have a substantial impact on the effectiveness of design results; this is because design processes include several decisions at each level (Levin, 1984). After gaining some knowledge of the issues, designers establish goals and then adhere to them. They generate several plans and exercise their judgement in a variety of areas, including information processing (Schmitt, 1999), strategy application (Almendra and Christiaans, 2009), and material allocation (Scaletsky and Marques, 2009), while selecting from a variety of parameter sets.

Apart from challenges encountered during the design process that were foreseeable or under the participants' control, some participants encountered difficulties throughout the design process as a result of events beyond their control or anticipation. The majority of these things were brought about by technical advancements. Some participants expressed anxiety over technical advancements such as sophisticated design tools and file format issues. One participant said that social events, such as news stories, would divert their focus away from the design process and cause them to become sluggish with their design project. One participant said that uncontrollable emotion is one of the things that might impede his design process.

The context in which 'rookie' design groups operated occasionally created problems in the design process that was beyond their control. The environmental variables of technological, social, cultural, and economic considerations, among others, may also influence designers' roles in the design process as well as their emotions. Among all of these elements, the most evident was the effect of technical considerations on the design process. The environmental elements stated by study participants were comparable to Ho's 'external factors of design process' (2010). External environmental changes would have an effect on designers' emotional experiences.

5.2.5   FUNCTIONS OF EMOTIONS IN THE DESIGN PROCESS

Participants in Empirical Study 2 suggested that emotions have varying effects on decision-making at various stages of the design process and that emotion works in tandem with the design process. The sharing described above is an illustration of this: One participant stated that he wished to alter his emotions via reading books or watching films if he encountered problems or bad feelings in daily life or even during his design research. He generally felt better following such entrainments and made better judgements. From a design viewpoint, a design student's emotion may be altered by events during the design process to better control his or her mood. Emotion is present at every level of the design process, albeit it varies according to stage. This notion emphasises that emotion would be altered throughout the design process; for example, negative feelings would be converted to positive ones via the use of specific techniques. This is an intriguing notion that warrants additional investigation. The notion that emotion pervades all stages of the design process is comparable to Kaufmann's (2003) assertion that designers make decisions based on their emotions. Kaufmann also emphasised the need to work with our emotions to enhance our ability to analyse received information. Because information analysis is a crucial activity for decision-making, the capacity to make judgements based on relevant information is increased. According to several participants, emotion not only played a role in every stage of the design process but also served as an assessment tool. Emotions might be detected in their reflections and work, influencing their decision-making the next time they encounter a similar design circumstance. In other words, emotions can function as a mirror for the designer and assist him or her in evaluating items and situations (Creusen, 1998). Alternatively, several participants stated that emotion may act as a catalyst, impairing or enhancing inexperienced design students' decision-making abilities. The sharing that was discussed previously demonstrates how emotion acts as a catalyst; as stated, one person listens to music. In other words, emotions may be aroused. To prepare inexperienced designer groups for the design process, they would keep their emotions in check by listening to music, for example. On the idea that positive emotion facilitates design decision-making during the design process, some aspects of an inexperienced design student's emotion may be increased. As a consequence, an emotional design approach would enhance the quality control of design results (Best, 2006). Emotions may be included in the design process in two ways: as a catalyst or as an integral part of the process. The various functions described by participants were similar to those described by Creusen (1998); emotion influences human behaviour, including information processing and decision-making, because emotion is not only a response to external and internal stimuli but also performs functions such as system regulation, action preparation and direction, communication of reaction and behavioural intention, and monitoring of an organism's internal state. Apart from challenges encountered during the design process that are foreseeable or under the participants' control, some participants encountered difficulties throughout the design process as a result of events beyond their control or anticipation. The majority of these challenges were brought about by technical advancements. Some participants expressed anxiety over technical advancements such as sophisticated design tools and file format issues. One participant said that social events, such as news stories, would divert their focus away from the design process and cause them to become sluggish with their design project. One participant said that uncontrollable emotion is one of the things that might impede their design process.

The feelings of anxiety, frustration and helplessness caused by environmental changes during the design process can be significantly detrimental to inexperienced design groups. As designers are often required to tackle complex challenges and make quick decisions, the lack of experience can be debilitating, leading to mental fatigue, decreased productivity and ineffective solutions. Furthermore, inexperienced designers may lack the resilience to cope with environmental changes, and may find themselves struggling to keep up with the pace of the design process. This could lead to a feeling of hopelessness and fear of failure, which can lead to a lack of motivation and the inability to be creative.

The effects of environmental changes on inexperienced design groups can be further exacerbated by the lack of resources, inadequate support, and limited guidance. Without access to the necessary resources, inexperienced designers may be unable to keep up with the pace of the design process and may find themselves overwhelmed and unable to make decisions effectively. Additionally, the lack of support and guidance can also lead to a lack of direction, resulting in ineffective solutions.

Overall, the effects of environmental changes on inexperienced design groups can be far-reaching and have significant impacts on the design process. Designers should be aware of the environmental factors that can affect their performance, and should be equipped with the skills and resources to cope with the challenges of the design process. Furthermore, experienced designers should provide guidance and support to inexperienced groups to ensure that they are able to effectively navigate the design process and achieve successful outcomes.

5.2.6 Emotion Management in the Design Process

Since emotions are subjective, various individuals may feel differently about the challenges encountered during the design process (Desmet, 2002). Emotions have a varying effect on people's decision-making abilities. The findings of the two empirical investigations propose certain techniques for the regulation of emotions during the process of design. Despite the participants' assertion that emotions appear to be disconnected from design studies, they did devise certain techniques for regulating their emotions during their design procedures. Certain participants employed self-regulation strategies to modify their emotions, such as engaging in activities like browsing the internet, viewing television, or engaging in sleep. According to the individual's account, the utilisation of these techniques resulted in a transformation of their negative or positive emotions into predominantly positive emotions. This emotional shift was observed to have a positive impact on their transferable skills,[1] specifically their ability to deliver presentations and make decisions in their daily lives.

According to the findings of Empirical Study 2, it was suggested by the participants that effective decision-making in design processes requires design students to regulate their emotions. The individual expressed their intention to employ certain techniques, such as commencing their day with music, as a means of priming themselves for effective decision-making in their academic pursuits. Desmet (2002) posited a comparable notion that emotions are individual responses to particular occurrences. Diverse individuals may exhibit varying emotional reactions towards challenges encountered during the design process, with some individuals experiencing favourable affective states while others may experience unfavourable affective states. According to the participants, the experience of positive emotions could potentially broaden their scope of consideration when it comes to addressing design problems. However, the presence of negative emotions may impede their ability to make effective design decisions due to increased stress levels. In general, the respondents posited that affect can exert both favourable and unfavourable effects on the process of designing. Drawing from their practical experience in design, the author posited that the incorporation of positive emotions could serve as a valuable adjunct to problem-solving in the design process. Moreover, the author suggested that acquiring the ability to maintain positive emotions may enhance creativity. The belief held was that experiencing negative emotions could impede accurate information processing, as it may result in an overestimation of the perceived threat and an underestimation of one's own ability. The individuals in question additionally proposed that they may require instruction on the effective regulation of their emotional responses.

The impact of emotion on diverse everyday activities, including learning, was investigated by Scherer and Tran (2001). The findings of their research have potential relevance to various aspects of the design education process, including but not limited to the preparedness to acquire knowledge, acquisition and assimilation of novel information, attribution of meaning, retention, application and extrapolation, and inclination towards replication. Scherer and Tran conducted an investigation into the impact of five primary categories of emotion on the process of learning. The study revealed that specific positive affective states (referred to as approach emotions) such as interest, hope, joy, and anticipation, facilitated the process of learning by promoting exploration and creativity (i.e. transfer and generalisation). In addition, the experience of negative emotions such as irritation, anger, hate, and aggressiveness compelled the learner to surmount obstacles and consider alternative problem-solving approaches (Scherer and Tran, 2001). Aken (2005) posited that designers with extensive experience and heightened emotional intelligence are better equipped than others to navigate the design process, thereby mitigating the risk of unmanaged process design and its attendant coordination and temporal challenges. Nevertheless, certain investigations on the intersection of design and emotion offer divergent perspectives from the individuals involved. Flam (1993) posited that negative emotions function as an alerting mechanism, prompting individuals to reassess their normative inclinations. In other words, when an individual experiences fear, their decision to engage in a particular behaviour is determined by a cost–benefit analysis (i.e. to calculate and compare the benefits and costs of their actions). Fear, being a type of negative emotion, induces a rational thought process and leads to a reduction in the scope of subjective rational calculations and a reordering of desires.

According to Flam's research, negative emotions can aid individuals in making rational and practical decisions by deterring them from taking unwarranted risks in situations where the potential loss outweighs the potential gain. Individuals who have encountered setbacks in the recent past tend to exhibit risk-averse behaviour. Furthermore, apprehension regarding failing again prompts individuals to engage in risk-averse actions instead of risk-taking actions. These individuals rely heavily on the regulations established by their respective organisations when making decisions (Flam, 1993). The findings of the focus group discussion indicate that the participants exhibited a lack of recognition regarding the potential advantages of negative emotions, as their focus was primarily the benefits of positive emotions. This implies a lack of comprehensive comprehension regarding the role of positive and negative emotions in the process of decision-making. Despite the participants' lack of awareness regarding the impact of emotions on their design studies, they acknowledged the potential influence of emotions on their daily lives, particularly on their transferable skills, such as coordination and presentation skills. Consequently, they endeavoured to regulate their emotions by employing self-discovered techniques. Scherer and Tran (2001) discovered that emotions are a significant factor in enhancing design outcomes, as they can bolster designers' exploration and creativity. Overall, the aforementioned results suggest that the exploration of design and emotion has the potential to enhance the design process for inexperienced design students with regard to their ability to make informed decisions. Currently, there is a lack of established avenues for the advancement of design and emotion education for inexperienced design students within the Hong Kong design education sector.

According to the respondents in Empirical Study 2, inexperienced design students/beginning designers must effectively regulate their emotions to make good design judgements. They said that they would begin using strategies such as listening to music to prepare for daily life decisions. Additionally, some of them would try various means to alter their emotions, such as 'surfing the internet', watching television, or sleeping, to improve their capacity to manage their design processes. According to Aken (2005), designers with greater experience and stronger emotions than others would handle the design process more effectively. They would avert the risk of an uncontrolled process design, which might result in coordination and scheduling issues.

The above satiation reflected that managing emotion is a critical aspect of the design process. It involves understanding the feelings and motivations of the user and incorporating those insights into the design. This can include understanding the user's values, beliefs, and preferences, as well as their emotional response to the design. It also involves creating a design that is emotionally appropriate and appealing. This means creating something that will evoke the desired emotions in the user. The goal of managing emotion is to create a design outcome that resonates with the user emotionally so that they will use it and enjoy it. Emotion Management is a key factor to consider when creating a successful design.

The Relationship of Emotion, Design Process, and Design Outcome

Participants stated that emotions will have an effect on design processes in both positive and negative ways. In light of their expertise with design methods, they stated that positive feelings might be beneficial as a supplemental tool for resolving design-related difficulties. Developing the ability to maintain a happy feeling may enable individuals to be even more creative. Additionally, they said that they may benefit from some instruction on how to regulate their emotions. Simultaneously, some participants stated that emotion would have an effect on the message conveyed by the goals and results. When they experience a pleasant mood, they may wish to create something that will convey positive messages. Their idea is comparable to that of Ben-Peshat (2004), in that design results influenced by users' emotional shifts improve the users' experience while also establishing a tight link between designers and users/audience. The designers would utilise goods to affect consumers' emotional responses and therefore increase user happiness (Tan, 1999). However, if they are experiencing negative emotions throughout the design process, they may be unable to work on projects that involve the dissemination of good messages or may even leave the design project altogether. Additionally, their experience indicated that they are frequently inspired by amusing things in their immediate environment as a beginning point for their creative process. Personal emotion plays a significant role in the development of a designer's own style, and Carrie Chau is no exception. The procedure by which Wun Ying created her Wun Ying Collection (Chau, 2007) is a typical illustration of this concept. similar to her designs, the inexperienced design students/ designers at the entry level would also transmit positive messages of happiness and love. Hence, Wun Ying Collection would be an example to demonstrate how emotion is involved in the design process and design outcomes.

According to the participants, their design decisions are informed by their personal experiences and individual preferences. Given that emotions are implicated in these experiences and individual preferences, it is likely that they exerted an influence on the design outcomes of the participants. The participants realised that emotions can aid designers in cultivating their individual artistic style. Designers tend to experience a heightened sense of enthusiasm and receptiveness towards exploring diverse methods to accomplish their design objectives when they integrate their preferred techniques and components into the design process. The acknowledgement of the significance of emotion in design parallels the outcome of research conducted by Denton et al. (2004) regarding the perspective of design students on design and emotion. Denton and his team discovered that the stimuli surrounding designers and designers' preconceptions have an impact on their design outcomes. Overall, the study's participants came to the realisation that emotions can serve as a valuable tool for novice design students in broadening their cognitive processes for effective problem-solving.

Emotion can be defined as a set of reactions that occur in response to external stimuli as well as a cognitive process that involves the evaluation of stimuli and situations (Scherer, 1984). Consequently, the incorporation of emotional considerations by designers during the decision-making process may have an impact on their ability to make decisions. Ho (2010) formulated the E-Wheel model by utilising Scherer's emotion concept to elucidate the interconnections among designers, decision-making in the design process, the external environment, and emotions. The emotional state of a designer may be influenced by the external environment, encompassing factors such as the social, cultural, and technological milieu. Simultaneously, designers undertake multiple decisions, such as information processing and material allocation, that incorporate their emotional state during the design process. Emotion has the potential to impact the internal factors that drive alterations in design outcomes. Designers' decision-making can be directly influenced by the external environment (without the involvement of their emotions). Therefore, the decisions made during the design process have the potential to impact the internal factors involved and subsequently shape the final design outcome. The feedback provided by the participants regarding the correlation between the external environment and emotions aligns with the notion put forth by Forlizzi, Disalvo, and Hannington (2003) that alterations in the external environment can impact the reflective emotional responses of designers, also referred to as 'emotional experience'. The impact of the external environment on a designer's emotions is primarily contingent upon the designer's level of awareness and introspection regarding the external environment (i.e. the external factors of the design process). Incorporating emotional considerations into the design process could potentially augment the decision-making capacity of the designer. Vosburg (1998) posited that designers' emotional changes prompt them to explore various directions, ultimately enhancing the calibre of their concepts. This assertion is corroborated by the present argument. The ability to discern between various types of information can be facilitated by emotional changes, thereby empowering designers to make optimal decisions through the employment of appropriate strategies.

The design process encompasses various forms of decision-making. Designers' emotional considerations influence their decision-making, which consequently impacts the design process. The point is illustrated by the suggestion offered by the participants of the focus group that the stages of information processing in the design process can elicit positive emotions. The heightened level of excitement experienced by a designer may potentially augment their capacity to analyse gathered information, thereby bolstering their decision-making abilities. Given that information processing is a crucial stage within the design process, the augmentation of this cognitive function through the incorporation of emotional factors would serve to bolster the designer's capacity to effectively navigate and manipulate the design process as a whole. In other words, optimising the design outcome would entail implementing the corresponding changes in the design process. Best (2006) posited that incorporating emotions into design processes can serve as a means of ensuring the quality of the design outcome.

Relationship between Decision-making and the Design Process

The process of design is commonly perceived as a means of making decisions (Levin, 1984). During the design process, designers establish objectives and subsequently generate multiple plans, utilising their discretion to select from various parameter sets. During the decision-making process, designers take into account various types of information, including practical considerations, relevant knowledge, and personal experience. Designers engage in information processing by employing various techniques such as deriving solutions, conducting consistency testing, and performing comparison and selection while adhering to time and budget constraints. In recent years, the study of decision-making processes has incorporated management skills to enhance the quality of outcomes and optimise the decision-making process (Longueville, 2003). The author has selected specific management skills, such as time management and risk management, to cultivate a methodical approach to decision-making. Scaletsky and Marques (2009) established a correlation between the aforementioned concepts and the allocation of resources.[2] The proposition was put forth that designers ought to acquire the aptitude for material allocation and selection, given that these factors hold significant sway over the design process. Therefore, based on a logical viewpoint, multiple determinations are taken during the process of designing (e.g. information processing and the allocation of materials). The findings of both Empirical Study 1 and Empirical Study 2 indicate that the feedback provided by the participants suggested a lack of comprehension regarding the connections between emotion and decision-making in the context of their design studies. The aforementioned studies have not yet been introduced to inexperienced design students.

Relationship between Emotion, Decision-making and the Design Process

Based on the feedback provided by the students, it appears that identifying the influence of a designer's personal emotions on decision-making is a more straightforward task than recognising their impact on the design process. The present study conducted a comprehensive review of the existing literature to investigate the role of emotions in the process of decision-making. The impact of the E-Wheel concept on designers' reflective emotional responses (specifically, their emotional experience) was observed to be influenced by changes in the external environment, which refers to the external factors of the design process (Ho, 2010). The impact of external factors on the emotions of designers is contingent upon the level of awareness and introspection that the designer possesses with regard to their external surroundings. Moreover, designers' decision-making capacity is augmented when they incorporate their emotional considerations into the design process. The emotional fluctuations experienced by designers serve as a catalyst for divergent thinking, ultimately leading to an improvement in the calibre of their ideation (Vosburg, 1998). Designers can leverage emotional changes to distinguish between various types of information, allowing them to select optimal problem-solving strategies (i.e. decision-making processes).

Simultaneously, the distinct phases of the design procedure encompass diverse internal elements that impact the process of making decisions (e.g. information processing and the allocation of materials). Designers make decisions based on internal factors that influence the design process, which may include emotional considerations. Consequently, the influence of emotions on these internal factors can lead to varied design outcomes. External factors – namely, those that are outside of a particular system or entity – may also be considered.

External factors that are beyond the direct control of designers have the potential to significantly influence their decision-making processes (without the involvement of their emotions). Therefore, the aforementioned decisions would have an impact on the internal factors of design processes and consequently influence final design outcomes. The E-Wheel and 3E models were formulated through a comprehensive analysis of existing literature, with the aim of elucidating the interplay between designers, emotions, and the intrinsic factors (e.g. information processing and material allocation) and extrinsic factors (i.e. those beyond the direct purview of designers, such as technological, social, cultural, and economic factors) that shape the design process (Ho, 2010). The emotional state of designers can be influenced by external factors, leading to alterations in decision-making that impact internal factors and, subsequently, the design process. This notion offers a perspective on how designers can leverage their emotions to generate reactions that enhance their design methodologies. The feedback provided by the user corroborates Darwin's (2007) theoretical framework on the interplay between emotions, behaviour, and decision-making. According to this framework, emotional expressions are either adapted to the environment or elicited by external stimuli, such as events or objects, and are subsequently shaped through social behaviours, such as human communication. With regard to the feedback concerning the influence of emotions on the decision-making process of the participants, an individual reported implementing certain measures, such as listening to music, to regulate their emotions in preparation for the return to school. One of the participants reported that their daily decision-making process was influenced by their emotions. They further elaborated that they regulated their emotions by means of a certain strategy.

Designers are influenced by their engagement with literary or cinematic media. The provided feedback corroborates Scherer's (1984) theoretical framework of emotion as a product of the interplay between mood and behaviour from a cognitive standpoint. The participants' readiness to exhibit behavioural responses to stimuli encountered in their daily lives, such as books, films, and music, aligns with Scherer's conceptualisation. Despite limited familiarity with the concept of emotion, the majority of participants acknowledged its likely impact on their daily cognitive processes and actions.

5.2.7 Design and Emotion in Design Studies

The participants in Empirical Study 2 were unaware of the real links between emotion and design, despite the fact that the majority believed they existed. They pioneered ways to 'brace up' or alter their own feelings on a daily basis, although they were unaware that these approaches would be utilised in design procedures. This was until the researcher steered them into an in-depth discussion of their collective experience and expertise in emotion and design (for the details, please refer to the transcript). They reflected on their design experiences and realised that they had previously attempted to modify their own emotions to influence their decision-making during the design process. The participants attempted to moderate their emotions to make more successful design judgements, but they frequently failed. They were unable to pinpoint the causes of their failure due to a lack of understanding of emotion and design. They argued that by learning more about the links between emotion and design, they would be able to investigate the most effective techniques for enhancing their performance during the design process. As a result, all participants felt that the inclusion of emotion in their design research would be beneficial. Simultaneously, based on their experience of incorporating their favourite components into their designs or making design judgements based on their personal working preferences, it was recognised that designers' emotions may aid them in developing their personal styles. When individuals experiment with different approaches and aspects that they enjoy during the design process, they become more enthused and engaged in exploring new ways to accomplish the design projects' aims. As a result, it is recognised that emotions would aid individuals in diversifying their problem-solving thought paths. Their acceptance of the critical role of emotion in design research is comparable to that of Western educators in other studies. Emotions have been identified as a critical element of education because they influence learners' and instructors' decision-making, creativity, and problem-solving abilities. Numerous nations, such as the United States of America (Adams, 2003), Belgium (Holper, 1998), and others, have begun to emphasise emotion in their revised curricula or workshops.

Inadequate Knowledge of Design and Emotion

The insufficiency of discussion and training on the subject of design and emotion was believed by the participants of Empirical Study 2 to be the root cause of their perceived inadequate knowledge on the matter. The study's participants reported limited exposure to emotional aspects in design studies. It was posited that the impact of affect on learners' decision-making capacity is a crucial element in the design of educational research. The participants exhibited a limited comprehension of the influence of emotion on decision-making processes and possessed a cursory understanding of the interplay between emotion and design studies. Despite the fact that approximately 40% of the participants in Empirical Study 1 were familiar with at least one of the terms 'emotionalised design', 'emotional design', and 'emotion design', 36.7% of the participants were unable to identify whether they had encountered these terms, and 80.2% of the participants were unable to identify whether they had utilised emotionalised design, emotional design, or emotion design. The results suggest that inexperienced design students have yet to delve into the correlation between emotion and design studies and that their instructors have provided minimal guidance on this topic.

In the 1960s, scholars in the field of education initiated an inquiry into the role of emotions in the processes of teaching and learning. Research has been conducted on the correlation between students' emotional state and their academic performance. The proposition was made that emotion constitutes a crucial facet of education, given its impact on the creative and problem-solving proficiencies of learners. Vygotsky's (1978) cognitive teaching theory was designed to facilitate learners' comprehension of novel information by leveraging their prior knowledge and enabling them to adjust their existing cognitive framework to incorporate the new information. Vygotsky conducted research on the impact of learners' emotions on their academic performance and self-efficacy. He highlighted that the sense of contentment linked with accomplishing a task holds significant sway over the extent of ease experienced by the student. Consequently, it is imperative for educators to exercise caution in designing assignments to ensure that they align with the present proficiency level of each student. Vygotsky acknowledged that educators' positive reinforcement of learners' past achievements and future potential plays a significant role in enhancing learners' self-efficacy. Hence, the establishment of positive student–teacher relationships is deemed to enhance teaching outcomes. The research conducted by Gage and Berliner (1992) expanded upon Vygotsky's (1978) findings by examining the aptitude of individuals with varying degrees of anxiety. It was noted that individuals who experience negative emotions, such as anxiety, tend to exhibit lower levels of performance compared to those who do not experience such anxiety.

It has been discovered that mood states can impact the efficacy of information processing during the learning process. Hence, it is imperative for educators to acknowledge the emotional traits of learners and opt for appropriate pedagogical practices that optimise academic achievements. Dweck et al. (1993) conducted a study on the relationship between students and teachers. The authors posited that the emotional displays of educators have the potential to impact the academic outcomes of students. It was noted that educators have the ability to assist learners in overcoming external obstacles by effectively conveying the appropriate emotions. In addition, overt manifestations of dissatisfaction by instructors can elicit feelings of shame and vexation among students. Nevertheless, such adverse affective states may potentially enhance the learner's drive to achieve success. The impact of the emotional expressions of educators on teaching outcomes has been observed. According to Zimmerman's (2001) findings, educational experiences that are engaging can lead to a sense of personal fulfilment for learners. In his study report, he highlighted that students possessing self-assurance in their abilities exhibit the capacity to undertake challenges, endure setbacks and disappointment, and exercise judgement. Individuals who possess a high level of comfort in interpersonal situations and exhibit a willingness to pursue their needs without fearing criticism are deemed as exhibiting traits of bravery. In addition, they demonstrate proficiency in collaborating with peers and exhibit a willingness to actively seek guidance and resources from instructors. Therefore, it is recommended that educators carefully choose tasks that align with the current performance level of each individual learner in order to promote a sense of self-fulfilment and foster positive emotional experiences. As an illustration, a particular student may undertake specific assignments, while other students engage in alternative undertakings. In this scenario, the educator would emphasise the distinctive abilities of every student. In summary, the aforementioned results suggest that the affective domain has the potential to impact the cognitive and behavioural aspects of students' learning.

Johnson and Johnson (2005) proposed a pedagogical approach that draws from their peace education framework, which emphasises the significance of emotional support in teaching. Their teaching plan centres on the development and sustenance of cooperative systems as well as the resolution of conflicts between educators and learners, which involves making decisions about complex issues related to maintaining peace and eliciting positive emotional responses. The authors recommended that educators refrain from utilising competitive grading methods and instead allow pupils to showcase their individual competencies and creativity. The proposed approach would establish a connection between the affective reactions of educators and learners and pedagogical results, such as innovative thinking and effective judgement. Numerous learning and teaching theories that incorporate emotional considerations suggest that emotions constitute a crucial component of education.

The Role of Emotion in the Curricula[3] (Excluding Design Curricula) of Other Countries

Adams (2003) noted that the United States, along with Belgium (Holper, 1998) and other nations, has recognised the significance of incorporating emotion into their educational curricula. Adams (2003) asserts that the United States was the first nation to acknowledge the necessity for educational restructuring. During the 1980s, the Education Commission in the United States compiled a range of societal risk indicators, which encompassed subpar academic performance, elevated levels of functional illiteracy among adults, and a downward trend in achievement test scores. American educators sought to identify a proficient educational framework that would yield superior academic results. Lang, Bradley, and Cuthbert (1998) noted that multiple standards-based reform measures were eventually implemented in different states, such as the Texas Assessment of Academic Skills (in 1991), the Washington Assessment of Student Learning (in 1993), and the Massachusetts Comprehensive Assessment System (also in 1993). Lang, Bradley, and Cuthbert (1998) highlighted that the education reforms in the United States, as per Camus' (1996) proposition, should prioritise emotions and feelings as crucial considerations for all democratic societies. The education reforms implemented in the United States served as a source of inspiration for European educators, who subsequently initiated their own progressive education reforms during the 1990s. According to European educators, the purpose of education extends beyond imparting academic knowledge and encompasses the development of students' capacity to make meaningful contributions to society. The educators prioritised the incorporation of psychological phenomena such as emotions, beliefs, attitudes, and feelings of students into the educational process. Holper (1998) asserted that Belgian educators acknowledged the enduring impact of emotional education on individuals and its significance as a fundamental component of the curriculum. Several workshops on creative expression in drama, printing, photography, and other related fields were offered by numerous secondary schools in Belgium. The study revealed that the workshops facilitated learners in developing comprehensive career plans and improving their decision-making skills, thereby enabling them to effectively navigate the demands of job interviews. Zabrondin et al. (1998) reported that problem-solving and creativity skills were included in the fundamental psychological curriculum for primary and secondary schools in Russia as part of the Psychology for Adolescents course. The study investigated various forms of creative endeavours and conceptualised creativity as the generation of ideas that exert an impact on the surrounding milieu. The activities designed to enhance communication skills were found to facilitate the development of subjective creativity. The individualised creative processes were instructed based on the particular aptitude of each learner. Consequently, the learners attained proficiency in regulating their emotional states and exercising agency in regard to future decision-making. Kondoyianni, Short, and Sideri (1998) posited that emotional education aligns with the objectives and subject matter of aesthetic education within the primary curriculum of Greece. The objectives of the primary artistic endeavours, encompassing design, were pertinent to eliciting emotional responses in this instance. The students were taught to value their own abilities and those of their classmates. They were motivated to produce artworks that fulfilled their individual expressive desires. The students acquired knowledge and acknowledged art, encompassing design, as a vehicle for self-expression and ingenuity. Throughout the course of their experience, the individuals gained a heightened perception of their own identity, explored novel and diverse methods of interpersonal exchange, and cultivated an enhanced comprehension of human engagement and connections. The students demonstrated an ability to establish links between their educational encounters and the broader societal and cultural milieu. Overall, research has shown that emotional education fosters creativity and improves students' decision-making skills. Drawing from the realm of general education, it can be posited that emotions play a pivotal role in shaping the daily lives of students. Furthermore, it can be argued that emotions constitute a form of knowledge that has a significant impact on the diverse learning abilities of learners, including but not limited to their creativity and problem-solving skills. In certain Western nations, subjects pertaining to the manipulation of emotions and emotional literacy have been established as a means of augmenting one's capacity for learning. The utilisation of emotion has been observed in various disciplines ranging from elementary education to advanced academic pursuits. The implementation of emotion training facilitates students' self-awareness and fosters a sense of value regarding their own academic achievements. Moreover, it facilitates the acquisition of self-awareness, fosters self-esteem and social skills, and promotes a disposition towards innovative and apprecia-tive collaboration with peers. Education professionals have the chance to comprehend the significance of emotions in the realm of education. Therefore, despite the fact that the participants in the second empirical study lacked comprehension regarding the influence of emotions on their design studies as indicated by the feedback and outcomes, integrating emotions into their studies would be a suitable approach to attain the aforementioned advantages. By gaining self-awareness and emotional intelligence, students can develop the ability to make sound decisions. Therefore, the learning processes and design studies of the students would be improved, resulting in enhanced personal growth.

Some scholars have investigated the role of emotion in teaching, and some teachers have started to learn about emotion and apply it to design education. Pampliega and Marroquin (1998) stated that a teacher not only takes on the role of instructor and transmitter of information but also acts as a personal development tutor in terms of the emotional dimensions of education. From this perspective, the interactive com-munication between educators and students enhances the students' problem-solving abilities, creativity, and cognitive training. Hence, it is understood that emotional education is a cross-disciplinary subject that permeates all study disciplines and is closely linked with cognitive training (including problem-solving and creativity).

Learning Programmes in Western Countries which Focus on Emotion

Following the release of Goleman's (2004) publication regarding the correlation between social and emotional intelligence, there has been an increase in the imple-mentation of structured educational programmes that prioritise emotional matters. The notion of emotional intelligence originated from Darwin's theory of emotional expression for survival, which pertains to the capacity to observe and differentiate the sentiments and emotions of others (Salovey and Mayer, 1990). The impact of affect on the utilisation of cognitive techniques, such as those employed in the reso-lution of problems, is a recognised phenomenon. The notion of emotional intelli-gence has been extensively embraced in various fields, leading to the emergence of several associated concepts, including emotional literacy, critical emotional literacy, and emotional creativity. These topics are currently of great interest in the field of education in England (Spendlove, 2007). The concept of emotional intelligence has been incorporated into educational policy, signifying the integration of emotions in the primary and secondary curricula in England. Certain academic courses have incorporated design methodologies, such as problem-solving and creativity, as auxil-iary resources to facilitate students' comprehension of the notion of emotion and their ability to identify their own emotional fluctuations during the process of learning.

The cultivation of emotional literacy among students has been shown to augment their creative capacities and aptitude for making informed decisions. Fredrickson (2000) posited that emotional literacy is predicated on the cultivation of positive affect in educational settings. Positive affect has been shown to facilitate effective decision-making and enhance creative ideation among students. The educational field has incorporated these concerns into the design curriculum to augment students' problem-solving and creativity skills as a consequence of the reformed curricula. Consequently, students from Western cultures exhibit a greater degree of familiarity with the notion of emotion when compared to their counterparts from Hong Kong. In addition, students in Western educational institutions exhibit a greater capacity for comprehending their own emotional states, as they are prompted to engage in creativity-based exercises that focus on emotional awareness, which are not typically included in the formal curriculum.

The Potential Relationships between Emotion and Design Education

As per the preceding research on pedagogy and cognition, there exists a close association between emotion and education, with both factors exerting a reciprocal influence on one another. The aforementioned results suggest that the implementation of emotional education could potentially augment the aptitudes of students in the domains of creativity and decision-making. The study of design necessitates the consideration of decision-making and creativity as significant factors. The authors Roberts and Burgess (1973) underscored the significance of decision-making in design study, particularly in addressing specific problems. They highlighted the importance of making decisions through an investigative and individualised approach rather than relying solely on past accomplishments. The academic areas of inquiry encompass the precise delineation of a problem and the systematic exploration of viable solutions that are both logical and rational, while also satisfying personal criteria. The significance of decision-making and creativity in design research was also emphasised by Green (1974). Green (1974) posited that the field of design study is focused on cultivating a discerning comprehension of human needs and acquiring proficiency in assessing the degree to which a given need has been adequately and effectively fulfilled. During the course of identifying a problem or need, designers are required to make decisions and subsequently evaluate a proposed solution. According to Green's findings, there exists a close correlation between the design process and the fundamental process of creative education as well as our daily lives. The presence of creativity in decision-making is a well-established phenomenon that holds relevance across various disciplines within the realm of design studies. The findings of Empirical Study 2 suggest that the impact of emotion on students' decision-making and creativity could be significant, thereby highlighting the potential significance of emotion in the field of design studies. The study of design has traditionally focused on topics such as ergonomics, sustainability, and accessibility. However, there is now a growing recognition of the potential significance of emotions as a new area of concern in design studies. This is in line with the work of scholars such as Wickens and Hollands (2000), McLennan (2004), and Mace et al. (1996) who have previously explored these topics. Design instructors and learners may collaborate to develop methodologies for analysing and identifying suitable and innovative approaches to incorporating emotion into educational programmes. Further investigation is required to gain a clear understanding of the influence of emotions on design study. This entails exploring the concepts and related theories on emotions in the design process, including creativity and decision-making.

The inquiry into why the subject of design and emotion is absent from the curriculum of beginner design students is a worthwhile pursuit particularly in light of the extensive research conducted over the past decade on the correlation between emotion and design. According to the findings of Empirical Study 2, the majority of participants expressed the opinion that undergraduate design students ought to concentrate primarily on fundamental subjects, such as typography and drawing. Furthermore, there is currently a lack of a structured pedagogical framework for teaching design and emotion at the foundational level. The limited research on the impact of design and emotion on core design studies has resulted in an undergraduate design curriculum that does not adequately incorporate emotional considerations for novice design students. The primary focus of undergraduate level education in design in Hong Kong is on intended outcomes, as per outcome-based principles. The aforementioned principles prioritise tangible objectives and results of the educational encounter over the methods of design. The majority of design educators prioritise instructing subjects that facilitate their students' attainment of desired results as opposed to exploring experimental topics during the undergraduate educational experience. The study participants exhibited a lack of familiarity with the field of design and emotion.

The evolution of the design sector and the exploration of pedagogical theories in design instruction have progressively impacted the character of design education. The process of design is considered to be multifaceted and nuanced, which produces a diverse array of satisfactory results. The aforementioned principle ought to be applicable in the realm of design education (Foa and Kozak, 1986). The current focus of design studies is on logical procedures that facilitate the identification of design issues pertaining to the requirements of users. Hence, certain design educators have proposed that design is a straightforward reaction to the market (Yeomans, 1990). The objective of design is to create feasible, commercially viable, and cost-effective products that cater to the needs, preferences, values, and functional demands of consumers. Consequently, it is imperative that design education endeavours to equip students with the necessary skills and knowledge to effectively function in the professional realm.

The current state of design education places excessive emphasis on logical problem-solving techniques (Stumpf, 2001). Students adhere to structured learning models when engaging in design tasks and the design process in addition to related concepts such as fundamental techniques and methodologies. According to Stumpf's (2001) proposal, design learning encompasses not only rational problem-solving but also social processes, hypothesis testing, and experiential learning. The implementation of integrated experiential learning fosters an environment in which students are motivated to seek out and investigate alternative approaches within the realm of design education, resulting in a more comprehensive understanding of social dynamics. Thus, Stumpf urged students of design to create designs that are rooted in their personal experiences. The author's investigation into the learning styles and perceptions of design students revealed that specific 'frames', such as the objective of the design task, can facilitate students' comprehension of the design problem and enable them to convert the problem into a resolution. According to Evans and Sommerville's (2007) proposition, designers must possess the ability to comprehend the advancements in social, cultural, technological, and economic domains and subsequently identify the demand for novel products. Hence, it is imperative for design educators to acknowledge the alterations in technology, user anticipations, and consumer preferences and aspirations. Evans and Sommerville expressed concern regarding the optimisation of the design curriculum, with a focus on the discrete nature of design as an activity within the broader decision-making process. The design curriculum in Hong Kong prioritises the application of logical reasoning in order to achieve feasible objectives and concrete results. However, it appears to be inadequate in facilitating the development of design novices' ability to effectively manage their design processes or to provide them with guidance on how to initiate the process with ease and efficiency. A group of design educators in Hong Kong convened to deliberate on strategies to enhance the efficacy of design pedagogy and curriculum with respect to learning outcomes. Based on empirical research, it has been determined that the design curriculum in Hong Kong is inadequate in facilitating the development of effective design processes among students. The study's participants expressed their disapproval of an exclusively utilitarian approach to education that prioritises the acquisition of practical skills. The study's second empirical findings indicate that the participants expressed a desire to acquire further knowledge regarding the correlation between emotion and design. They believed that such knowledge would enable them to investigate methods to improve their decision-making abilities during the design process.

Thus, the study participants arrived at a collective agreement that integrating emotional considerations into their design research would yield benefits. Improving the quality of design education and research can be accomplished by increasing recognition among design educators and novice design students of 1) the relationship between design and emotion and 2) the resulting influence of emotion on design decision-making and procedures. Although numerous studies have investigated the effects of emotions on teaching and learning, a relatively small cohort of design educators recognises the influence of emotion on design education.

Several scholars originally from the Western hemisphere have suggested the incorporation of emotional education and creativity into the existing curriculum. According to the theory of emotional intelligence, the influence of emotions on the execution of learning strategies, such as the acquisition of implicit and innovative knowledge, is substantial (i.e. the knowledge that we acquire through experience). Salovey and Lopes proposed the incorporation of a social and emotional learning (SEL) programme into both the academic and non-academic aspects of schools' existing curricula (2004). The SEL programme aims to improve the social and emotional well-being of students by providing them with practical skills training, including decision-making skills. According to Salovey and Lopes (2004), the inclusion of decision-making skills is a vital component in emotional education methodologies. The triadic schema for art and design education was introduced by Spendlove in 2007. This schema was influenced by the recognition of emotion in the professional design field, which includes emotional design (Norman, 2004) and emotional ergonomics (Seymour and Powell, 2003). The schema in question is focused on the concepts of creativity, learning, and product orientation. The aim of the schema is to cultivate affective engagement among students in their artistic pursuits within the individual domain, facilitate emotional engagement through favourable and supportive learning contexts within the procedural domain, and enable students to take emotional responsibility for their creative outcomes (product domain). The enhanced emotional intelligence and favourable emotional state of students can have a substantial influence on their ability to make decisions and generate creative ideas. Spendlove, a scholar specialising in emotional literacy, augmented the art and design curricula at primary and secondary educational institutions with the objective of cultivating emotional proficiency among pupils. The proposal recommends offering emotional assistance and contextualisation to students during the personal and process stages. The emotional consequences that emerge in the personal stage can be amplified by integrating the emotional requisites of the recipient in the process phase. Incorporating emotional factors into the product domain has the potential to improve design outcomes.

The educational methodologies mentioned earlier propose that emotions can impact both the process of decision-making and the generation of creative output. Although design and emotion have gained widespread acceptance among educators in the Western hemisphere, it seems that design educators and novice design students in Hong Kong have not yet fully engaged with this field. The students in question face challenges in establishing a connection between their emotional state and the design process. Meanwhile, educators encounter obstacles in providing efficacious remedies to aid them in regulating their emotions throughout this process. The emotional state of students may hinder their ability to overcome challenges during the design process, which could have an impact on their performance and the overall effectiveness of the design process.

Potential Teaching Strategies in Design and Emotion Studies

Empirical evidence suggests that affective states play a crucial role in the process of learning and instruction. Several nations have incorporated educational activities that encompass emotional knowledge and training into their distinct curricula. The primary objective of these educational assignments is to augment the student's proficiencies in critical thinking and innovative ideation. Emotion has been incorporated into creative disciplines, such as art and photography, in various other Western educational programmes. The aim of these courses is to enhance the emotional awareness and expression of the students. The aforementioned educational programmes propose a correlation between instruction in art, creativity, photography, and other primary subjects within the field of design studies and instruction in emotional development. Empirical Study 2 investigated the correlation between learning and emotion. The present study offers initial findings regarding the role of emotion in the design processes of inexperienced design students. The findings indicate that the intersection of design and emotion holds promise as a prospective subject in the forthcoming design syllabus in Hong Kong. This prompts inquiry into the methodology of incorporating the topic of design and emotion within the realm of design studies.

The second empirical study's participants exhibited limited design experience within the duration of their brief design study. The majority of the respondents reported that they were susceptible to emotional arousal from their surroundings, including literature, cinema, and music. The individuals comprehended that the aforementioned emotional alterations could potentially impact their routine choices. The study's participants comprehended that the emotional transformations they experienced in their daily lives could be relevant to their design methodologies. They also devised distinct approaches to address the challenges encountered during the design process. Hence, comprehending the correlation between emotion and design, predicated on encountered emotions and individual experience, may serve as a catalyst for fledgling design students to take efficacious decisions and facilitate their adept manipulation of the design process. In this scenario, the learners would engage in active learning through personal experience rather than relying on the instruction of their educators (Erev and Haruvy, 2013; Schweitzer and Cachon, 2000). Hence the utilisation of personal experience has the potential to augment the proficiency of novice design students in managing the design process. Incidentally, the aforementioned investigation exposed the ineffectiveness of conventional pedagogical approaches, which commonly depend on the direction of instructors.

As per the respondents, there exists a disparity in the ability of inexperienced design students to make judicious decisions while undertaking the design process. Despite receiving valuable instruction from tutors on how to navigate various stages of the design process, certain students still encounter difficulties in comprehending the intricate nature of design development. Hence, it is recommended that a paradigm shift be implemented in the teaching strategies of design education to enable students to engage in self-exploration and enhance their intuitive learning capabilities.

Implanting Emotion in the Design Curriculum through Action Learning

Western educational curricula that incorporate emotions into the learning process are predicated on students' experiential learning and reflective analysis of authentic situations. The approach being discussed resembles the methodology employed in action learning as described by Kramer in 2008 (refer to the next paragraph for a detailed explanation of action learning). Thus, the implementation of action learning may serve as a means to incorporate affective elements into the design pedagogy.

Action learning is an educational methodology that involves students in addressing authentic problems and engaging in reflective practices to enhance their learning outcomes (Kramer, 2008). Through addressing tangible problems, students acquire practical knowledge and emotional intelligence, which they subsequently articulate in their reflective analyses. Subsequently, individuals can utilise their introspective analyses to address additional concerns. In this instance, the acquisition of knowledge and reflection by students occurs through practical application and experiential learning as opposed to conventional pedagogical methods. The significance of reflection is underscored throughout the action learning process. Collaboration is a common practice in which individuals work in small groups to facilitate self-reflection and evaluation of their actions. The act of reflecting serves as a means of acquiring knowledge and directing the actions of students towards improvement in their academic performance. The framework of action learning can be utilised for the purpose of research experience and the development of case studies. This methodology is specifically well suited for postgraduate design instruction (Chattaraman, Sankar, and Vallone, 2010). The study conducted by Chattaraman, Sankar, and Vallone revealed that the implementation of action learning can effectively equip graduate design students with the necessary skills and knowledge to excel in creative and management careers. This is achieved through the utilisation of action learning as a means of facilitating personal and professional growth as well as fostering self-assurance in the learner.

The original action learning model failed to account for the role of emotions. According to Vince and Martin's (1993) perspective, the incorporation of a theory of emotions in management learning is necessary to complement the action learning model. The proposed approach fundamentally redefines the role of ego defence mechanisms in the processes of learning and development. The argument posits that prioritising rationality overlooks the importance of emotions, thereby reinforcing the established political norms within organisational contexts. The authors posited that the phenomenon of personal transformation in action learning is attributable to the interplay of psychological and political processes, encompassing emotional and rational factors. Clarke and Roome (1999) identified that the action learning process is influenced by various components, including practices, work routines, individual job focus, and context. These elements impact an individual's emotional inclination and ultimately determine his or her willingness to persist with the learning process. Individuals who fail to surmount the challenges inherent in action learning may experience a decline in motivation to persist with the process. The presence of negative emotions such as 'lost motivation' and 'perceived lack of power' implies that emotions play a role in, and potentially even influence, the progression of the learning process.

Drawing from the aforementioned theories, the action learning methodology has been implemented in various extant design curricula as a means of reconciling the dichotomy between theoretical concepts and practical application (Li and Barrett, 2002). Furthermore, as previously discussed, research has demonstrated that emotions have a significant impact on decision-making processes and creative outcomes within the context of design education. Thus, based on the available evidence, it appears that the incorporation of emotion into the design curriculum can be facilitated through the implementation of action learning.

5.3 LIMITATIONS

The study yielded intriguing findings regarding the correlation between emotion and design with respect to decision-making and the comprehension of the design and emotion concept among inexperienced design students. These discoveries contributed to a comprehensive understanding of the aforementioned relationship. However, the present study is subject to several limitations pertaining to the existing literature on design and emotion as well as the employed research methodologies. These limitations are elaborated upon in the subsequent sections.

Limitations of the Study Sample

Due to resource constraints, all respondents were Hong Kong students. Salili and Hoosain (2001) assert that Hong Kong pupils are less self-sufficient in their learning and rely substantially on their professors. Additionally, teachers prepare the majority of their educational experiences. They are not like Western pupils, who are encouraged to develop their own learning processes and outcomes. Students in Hong Kong and Western countries have significantly distinct learning methods; Hong Kong students place a higher premium on skill development. They engage in less self-exploration or reflection on their experiences, which is common among inexperienced design students/designers at the entry level. The passive learning style of the majority of Hong Kong inexperienced design students/designers at the basic level explains in part why they do not have as much knowledge of emotion in design as inexperienced design students at comparable educational levels elsewhere. Compared to elsewhere, fewer inexperienced Hong Kong design students/designers at the entry level investigated the role of emotion in their design decision-making. Although they may have experimented with strategies to manage their emotions for more efficient decision-making, they have never self-examined the reasons behind their inability to control their emotions. They require support from professors to investigate their challenges with the design process and the factors that contribute to their lack of knowledge of emotion and design. In comparison to Western pupils, Hong Kong students clearly lack a grasp of the emotional impact of design. However, as design education continues to globalise, Hong Kong design education is incorporating more Westernised educational approaches (Sutherland, 2002). Due to the growing popularity of cross-cultural studies in Hong Kong design education, multicultural design teaching and learning approaches are being used (Biggs and Watkins, 2001). The divide between Hong Kong and the West in terms of design education is thus decreasing. Accordingly, additional research is required to determine the difference in knowledge between Hong Kong and Western pupils. Another constraint was the small sample size of this study. Due to resource constraints, only eight people were enlisted.

Difficulties in Scrutinising the Design Practice of Novice Design Groups

Due to resource constraints, an action study on how young designers deal with emotions during their decision-making and design processes was not possible. Action research, in which people are invited to collaborate on a design project, may assist researchers in identifying the obstacles encountered by inexperienced design students and their emotional shifts during the design process. While action research is an appropriate mechanism for observing participants' performance, rather than critically evaluating and making comparisons of their understanding of emotion, decision-making and the design process, it cannot provide information about how the student recognises the role of emotion and its significance in the design process.

5.3.1 PRELIMINARY INVESTIGATION OF EMOTION IN THE DESIGN EDUCATION

This research is a preliminary inquiry into how the design process is impacted by emotion, from the viewpoint of inexperienced design students. The study makes use of recent writings on designers, design processes, and emotions. To fully comprehend the connection between design and emotion, the researcher investigated and evaluated the key connected papers. Several words from the earlier research, including 'emotionalised design', 'emotional design', and 'emotion design' were found and reinterpreted. In recent years, the theories and ideas around design and emotion have matured and grown more organised. A few recently established ideas in design and emotion may not be able to be completely studied owing to availability and accessibility issues despite the fact that the bulk of topics connected to design and emotion, such as the result, users/consumers, and designers, were evaluated. Although the most recent and accessible research was used in this study to evaluate these recently formed notions, the effects on the overall research findings were minimal.

5.3.2 LIMITATIONS OF APPLYING DIFFERENT RESEARCH METHODS TO DESIGN STUDIES

The present literature review scrutinises diverse viewpoints on the intersection of design and emotion and reveals a dearth of research that delves into this intersection from the vantage point of inexperienced design students. Hence it is imperative to conduct additional research to acquire a profound and all-encompassing comprehension of the emotional experiences of novice design students during their decision-making and design processes. An action research methodology could be utilised to explore the correlations between students' comprehension of emotions and their challenges in the process of design. Action research is a methodology that entails active involvement in organisational change. It encompasses reflective processes of progressive problem-solving, which are carried out by individuals working collaboratively in teams within a real-life context (Somekh and Noffke, 2009). One potential approach to engaging students in design projects involves facilitating their participation while under observation, with the entire process being recorded for later analysis. The utilisation of the action research methodology has the potential to generate novel empirical findings pertaining to the manifestation of emotions. Notwithstanding this, it could be argued that this approach may be insufficient in terms of adequately elucidating the challenges that students encounter during their design processes as well as the resultant alterations in their physical responses, such as gestures, eye movement tracking,[4] facial expressions,[5] heartbeat,[6] skin response,[7] brainwave response, and other subtle manifestations that may not be readily discernible. If such a scenario were to occur, the outcomes of an action research investigation would not provide satisfactory responses to the research inquiries of this particular study. The present investigation employed a mixed research methodology that incorporated both quantitative and qualitative data collection and analysis techniques (i.e. questionnaires and a focus group). The focus group participants were prompted to inquire and bring up topics pertaining to the intersection of design and emotion. The study comprehended the challenges that triggered alterations in their affective states and the interconnections between affect and decision-making within their design procedures based on the feedback obtained from their focus group.

5.3.3 Unpredictable Responses from the Participants Leading to Limited Results

The primary objective of the three inquiries in Empirical Study 1 was to determine the participants' level of familiarity with the three fundamental terms, namely 'emotionalised design', 'emotional design', and 'emotion design'. The three inquiries presented in the later part of Empirical Study 1 functioned as subsequent inquiries aimed at exploring the participants' comprehension of the meanings and concepts associated with the terms 'emotionalised design', 'emotional design', and 'emotion design'. The majority of the participants exhibited unfamiliarity with the aforementioned three terms, which constituted an unforeseen discovery. In Empirical Study 1, it was found that 40% of participants who were familiar with the terms 'emotionalised design', 'emotional design', and 'emotion design' were able to provide indicative responses regarding their understanding of these terms. A limited number of respondents were familiar with the aforementioned trio of emotion-related concepts due to several factors. The discourse on emotion has not yet achieved widespread attention within the realm of design education. Consequently, the individuals involved in the study were not familiar with the aforementioned associated concepts. The notion of design and emotion in design research is predominantly deliberated at a sophisticated level during conferences and seminars. Thus it is understandable that the respondents were unfamiliar with the aforementioned terms given that they were still in the rudimentary stage.

5.3.4 Generalising the Research Results to Other Countries

There may be inquiries regarding the potential impact of restricting the sample to solely Hong Kong students on the statistical significance of the research findings. The present study investigated the potential variations that may arise when extrapolating the findings to inexperienced design students in different nations.

5.4 CHANGES IN HONG KONG EDUCATION TO ALIGN WITH THE INTERNATIONAL LEARNING STYLE

5.4.1 Educational Reform

Hong Kong is a globally recognised metropolis that maintains strong international affiliations with other nations. In contemporary times, there has been a growing momentum towards economic and social progress, driven by the imperative of national sustenance and expansion in global and local domains. The Hong Kong administration has implemented diverse educational and social measures with the aim of enhancing the competitiveness of Hong Kong vis-à-vis other East Asian nations. Higher education is regarded as a crucial element in enhancing the competitiveness of Hong Kong. The higher education system in Hong Kong was reviewed by the University Grants

Committee[8] in 1996 in response to the government's education policy (UGC, 2001). This analysis considered the policy objectives of the government with respect to the present dearth of opportunities for post-secondary education. The reform encompassed significant facets of higher education provision, comprising an administrative structure for the governance of universities. The study also scrutinised matters pertaining to the delineation of higher education, the function of higher education, the framework of governance for the higher education domain, university administration, research, and the identification of variables that impacted the progression of higher education in Hong Kong (UGC, 2001). After conducting an extensive evaluation, the educational programme for tertiary studies, including undergraduate studies, in Hong Kong underwent revision across all fields of study. Since 1996, Hong Kong has progressively incorporated Western pedagogical approaches. The current undergraduate curriculum in Hong Kong has shifted its emphasis from memory and skills to studio-based[9] practice.

Technological Development to Access Worldwide Information

The emergence of technology has led to the proliferation of novel media platforms, including the internet and mobile devices, that enable students to conveniently obtain information from around the globe. Hong Kong students exhibit similar patterns of interaction with the external environment as their Western counterparts. The convenient accessibility of information enables inexperienced design students in Hong Kong to conveniently explore and peruse various subjects, thereby motivating them to embrace a global design approach. The internet serves as a platform that facilitates the acquisition and assimilation of new knowledge with ease. Students have the agency to proactively investigate subjects that they find intriguing and significant. The advancement of technology has facilitated the emergence of a universal approach to learning in Hong Kong.

5.4.2 CHANGES IN THE PASSIVE LEARNING STYLE OF HONG KONG

The Learning Style of Hong Kong Students in the Past

According to certain scholars, it has been proposed that students in Hong Kong tend to adopt a passive approach to learning (Biggs and Watkins, 2001). Despite being adept at examinations and prioritising grades, Hong Kong students tend to attain their educational objectives primarily through rote memorisation. In contrast to their Western counterparts, students in Hong Kong tend to employ a direct and focused approach to problem-solving, characterised by a linear thought process. The students exhibit infrequent utilisation of personal initiative in investigating subject matter beyond the scope of their prescribed academic coursework. According to the research conducted by Salili and Hoosain (2001), it was observed that students in Hong Kong exhibit a low degree of independence in their learning and tend to depend heavily on their educators for the majority of their educational experiences. This form of instruction is a discerning educational encounter orchestrated by instructors to furnish pragmatic opportunities for resolving problems that facilitate pupils' advancement. Thus it can be observed that Hong Kong students have traditionally engaged in a relatively low degree of self-exploration throughout their educational journeys.

Adopting the Western Learning Style

Hong Kong educators have recently incorporated Western educational techniques that prioritise reflection and view learning as a restructuring process, thereby shifting away from the conventional passive learning approach. This approach grants learners greater control and responsibility over their learning experience (Horton, 1971). This methodology not only enhances the learners' motivation to learn and fosters their perseverance in the learning process but also facilitates their comprehension of the subject matter by prompting them to reflect on each stage of the learning process. In contrast to a passive learning approach, the current teaching and learning practices in Hong Kong prioritise the establishment of a balanced dialogue between educators and learners (Freire, 1974). The educators facilitate the learners in comprehending their own selves and the surrounding milieu (Paterson, 1979).

There are Western scholars who maintain the belief that students in Hong Kong persist in being passive learners (Biggs and Watkins, 2001). The current educational reforms and advancements in technology are motivating students to access global information. Consequently, students in Hong Kong are undergoing a rise in participatory learning and heightened levels of self-discovery. In general, students in Hong Kong are increasingly embracing an active learning approach. Therefore, it can be inferred that the impact of the passive learning approach was comparatively insignificant.

The Generalisability of the Results to Act as a Reference for Other Countries

As mentioned, some scholars have argued that the passive learning style of some students has caused them to be comparatively less aware of the influence of emotion. However, as indicated by some education reform reports, Hong Kong has recently undergone a significant educational change (UGC, 2001). The influence of education reforms and technological development means that many of the participants in this study were able to explore their surrounding environments on their own and adapt to the Western active learning style. Therefore, the results of this study are generalisable to novice design students worldwide, and may also serve as a reference in other countries.

To explore the relationship between emotion and design in terms of decision-making, it was necessary to understand the role of emotions in the design process, investigate how emotions affect decision-making and explore the participants' understanding of the concept of design and emotion. The literature review explored the theoretical understandings of the role of emotions in the design process and how emotions affect decision-making. A number of similar terms and concepts relating to the roles of designers, design outcomes, and users/consumers in the design process were found. These meanings of these concepts were clarified using the newly developed E-Wheel and 3E models. Although the literature indicates that the relationship between design and emotion is well understood in the design research field, less research has investigated design and emotion from the perspective of novice design students in design processes, who mostly experience difficulty in manipulating the design process because of the influence of emotion.

Empirical Studies 1 and 2 investigated whether novice design students understand and are familiar with the topic of design and emotion. Empirical Study 1 explored the participants' perception and understanding of design and emotion how they experienced design and emotion in their design consumption and design processes. Empirical Study 2 corroborated the results from Empirical Study 1 to give a more in-depth understanding of the participants' understanding and experience of design and emotion. In all, the two empirical studies explored novice design students' experience and awareness of design and emotion and the role of emotions in decision-making and the design process. The main findings of the studies are as follows:

- Novice design students have an inadequate understanding of the particular terms relating to design and emotion, i.e. 'emotionalised design', 'emotional design', and 'emotion design'.
- Design processes that involve emotion improve the quality assurance of the design outcomes. Emotion can be involved in the design process in two ways: working as a catalyst and working along with the design process.
- As design and emotion was a new topic for the participants, the participants were not sure whether they had experienced the influence of emotional concerns in design.
- Few of the participants had experienced the influence of emotion in the design experience, where the users' consumption actions involve both the designer and the user.
- The two empirical studies produced some contradictory findings. Although the participants mentioned that emotion seems to have no relationship with design studies, they had investigated some methods of managing their emotions. Some of the participants had initiated methods to change their emotions by, for example, surfing the internet, watching TV and sleeping, to enhance their transferable skills, such as their presentation skills, to handle their design-making in their daily life.
- Although most of the participants had little knowledge of the concept of emotion, they agreed that emotion would probably affect their thinking and behaviour in their daily lives.
- The participants indicated that emotion had a short-term effect on their decision makings and that it also contributed to the design outcomes.
- The results of the empirical studies indicate that Hong Kong students have less understanding of the influence of emotional concerns on design than Western students.
- However, the participants in Empirical Study 2 were able to identify the emotional changes they experienced in their decision-making and the design process under the lead of the questions and the background information provided on emotion and design.
- As design and emotion was found to help novice design students to manipulate their design processes, it is suggested that the topic of design and emotion should be included in the design studies curriculum.

- This study recommends a paradigm shift in teaching strategies away from following instruction towards intuitive self-directed learning.
- The results of this study are generalisable to novice design students in all countries and the findings can be taken as a reference in other predominantly Chinese societies, such as China and Taiwan.

NOTES

1 Transferable skills are skills that can be used in different occupations, such as design, teaching, and medicine, regardless of the type of work undertaken (Bowes, 1999).
2 Resource allocation is the scheduling of activities and the resources required by those activities while taking into consideration both the resource availability and project time (Scaletsky and Marques, 2009).
3 Curricula refer to the sets of courses (and their contents) that are offered at schools and universities.
4 Eye movement tracking is used to measure eye positions and movement.
5 A facial expression analysis tool categorises facial expressions into a coding system that allows the coding of any facial action in terms of the smallest visible unit of muscular activity (action units).
6 A heart rate monitor is used to record heart rate changes over short periods during which the heart rate may change.
7 Skin and brainwave responses are measured by an IBM emotion mouse. These measurements are difficult for the subject to control.
8 The University Grants Committee (UGC) of Hong Kong is a non-statutory advisory committee responsible for advising the HKSAR government on the development and funding needs of higher education institutions.
9 Studio-based refers to the practice where students learn by sharing a studio and exchanging ideas with fellow students.

6 Conclusion

6.1 DESIGN AND EMOTION STUDY FOR NOVICE DESIGN GROUPS

The purpose of this research was to examine the effect of emotions on decision-making and the design process that inexperienced design students/designers at the entry level may confront. Rather than examining how extrinsic information assists novice design groups in managing the development process and creative approach, the relationship between emotions and design processes, as well as how students' emotions varied in response to their decision-making, were examined. How did 'design and emotion' notions affect younger designers and even inexperienced design students? How did they approach their design research with an understanding of the importance of emotions?

Two empirical investigations were conducted to ascertain how designers at the entry level viewed the relevance of emotion and its roles in their design processes. In the first, a questionnaire was used to elicit broad knowledge about the links between emotion and design among inexperienced design students/designers at the entry level. Based on the results of this empirical study, an in-depth focus group was conducted to clarify the data and to gain a more detailed understanding of the relationships between students' emotions and their design processes. The related issues, such as how emotion affected the decision-making of inexperienced design students/designers at the entry level, as well as the corresponding design outcomes, were also investigated.

In the 1990s, in response to selected psychological results in the literature review, the design professions began to investigate the link between design and emotion. According to some researchers, design may stimulate user/consumer emotions; according to others, design incorporates emotion, which enhances relationships between designers and users. Several comparable concepts connected to emotional design were invoked in these study investigations, including emotion design, emotive design, and emotionalise design. In the majority of cases, these words were applied to the root of English grammar, although some researchers used the phrase as a general term to refer to anything involving emotion. There was a lack of clarity on the precise meaning of these phrases and how they should be employed in design and emotion research. To date, some academics have classified studies based on their study

DOI: 10.1201/9781003388920-6

interests, the differentiations between these classifications and the notions associated with these terminologies. Additionally, the primary functions of these three terms were identified, that is, designers, design output, and users/consumers throughout the design cycle of conception, production, and reflection.

As demonstrated by the results of the two empirical investigations, inexperienced design students/designers at the entry level had an insufficient grasp of design and emotion. Few participants were familiar with the words associated with the umbrella phrase 'design and emotion'. Few of them were familiar with the phrases 'emotion design', 'emotional design', and 'emotionalise design'. The majority of pupils were unable to identify appropriate subjects with specific phrases. They were unaware of the connections between emotion and designers, users/consumers, and design outputs. Indeed, they lacked a fundamental understanding of emotion. They were perplexed by the origins of emotion and struggled to establish an adequate definition.

6.2   RECOGNISED RELATIONSHIPS BETWEEN EMOTION AND DECISION-MAKING

Students typically realise that emotion affects their daily decision-making, but many are unaware of how this happens. They control their own emotions to prepare for daily life decisions through activities, such as listening to music, reading books, and watching films. The majority of inexperienced design students/designers at the entry level like to experience happy emotions, since they believe that pleasant emotions should aid in decision-making. In other words, unpleasant emotions can hamper designers' decision-making abilities and create challenges during the design process. These issues include ineffective time management, a lack of knowledge, ineffective risk management, ineffective communication with team members/tutors, and ineffective emotion management. Apart from obstacles in the design process that are foreseeable or under the participants' control, other difficulties in the design process are a result of variables beyond their control. These are the obstacles imposed on design groups by the environment, and they are beyond the novice design groups' control. Environmental elements, such as technological, social, cultural, and economic considerations all have an effect on the role and feelings of designers during the design process.

6.3   RECOGNISED ROLES OF EMOTION IN THE DESIGN PROCESS

Following an investigation of the links between emotions and decision-making skills, many functions for emotions in the design process were found. As per novice design groups' experiences, because emotion may boost the decision-making capacity of design students/designers at the entry level, decision-making functions at all stages of the design process will be impacted. At certain phases of the design process, the emotion will also be altered; for example, negative feelings may be transformed into positive ones. Design processes include several decisions, and the influence of emotion on decision-making may impact the success of design solutions, even if the effect is temporary. Not only does emotion play a role in every stage of the design process, but it also serves as an assessment tool when designers reflect on their emotional experiences. Additionally, emotion acts as a catalyst that may be activated and can hinder or boost designers' decision-making abilities, implying that an emotional design process can degrade or improve the quality assurance of design products. To prepare for the design process, designers at the entry level maintain control of their emotions by relaxing; for example, they might listen to music.

6.4   EMOTION MANAGEMENT AS A FACTOR IN DESIGN PROCESS MANAGEMENT

Because the growth of design studies and the expansion of design education are inextricably linked, it is critical to incorporate design and emotion into design studies. While some inexperienced design students/designers at the entry level realise that design procedures need a great deal of decision-making, few recognise the link between emotions and design processes. At the beginning of their careers, inexperienced design students/designers incorporated emotions in their decision-making processes, but they were unaware of the significance of emotions in design. They were exposed to the interplay between emotion and design and included notions related to these aspects in their design processes, but they lacked sufficient comprehensive knowledge on manipulating the design process. Some inexperienced design students/designers at the entry level attempted to initiate methods for resolving emotional difficulties in the design process on their own and discovered effective methods for managing the design process through trial and error; however, they continued to encounter difficulties in their design processes. Thus, by emphasising the significance of emotion in decision-making at each level of the design process, students may be able to improve their capacity to regulate their emotions, as well as their decision-making skills. They will then be able to successfully apply their design methodology.

6.5   POSSIBLE DIRECTIONS IN THE REFORM OF HONG KONG DESIGN STUDIES

In contrast to the majority of research on design and emotion, this preliminary design and emotion report focused on design studies for inexperienced design students/designers at the entry level in Hong Kong. Due to the conventional secondary educational system's emphasis on outcomes rather than process (Siu, 2003), inexperienced design students/designers at the entry level in Hong Kong are less active than Western students. In comparison to Western students/designers at the entry level, Hong Kong's junior design groups receive limited instruction in critical thinking. They seldom actively recollect their prior decision-making and difficulty-solving experiences throughout the design process or conduct more research into the reasons for the challenges encountered and the judgements made. Additionally, they seldom engage in subsequent talks that are validated by decision-making or unpleasant feelings (Brissaud, Garro, and Poveda, 2003). Thus, there is a need to investigate the study of design and emotion in general design lessons in Hong Kong design-degree studies to introduce a general understanding of the relationships between emotions and design processes and to raise the awareness of inexperienced design students/ designers at the entry level about the critical role of emotions in design through the design learning experience. As a result, inexperienced design students/designers at the entry level may gain an understanding of how emotions influence their decision-making and design processes, as well as research alternative techniques for facilitating their design processes. At the novice level, design students/designers comprehend emotional shifts and, accordingly, strive to enhance the quality and standard of design outputs to make them unique, innovative and interesting.

References

Adams, J. E. (2003). Education reform. In J. W. Guthrie (Ed.), *Encyclopedia of Education* (pp. 512–515). Macmillan Publishers.

Aken, J. E. V. (2005). Valid knowledge for the professional design of large and complex design processes. *Design Studies*, *26*(4), 379–404.

Akin, O. (1984). An exploration of the design process. *Design Methods and Theories*, *13*, 115–119.

Akin, O. (1998). Variants of design cognition. *Proceedings of the Knowing and Learning to Design Conference*. Georgia Institute of Technology.

Alexander, R. (2001). *Rinzen Presents RMX Extended Play*. Gestalten Verlag.

Almendra, R. and Christiaans, H. (2009). Improving design processes through better decision-making: An experiment with a decision-making support tool. *Proceedings of the International Association of Societies of Design Research*. Retrieved 26 January 2011, from www.iasdr2009.org/ap/Papers/Orally%20Presented%20Papers/Design%20Method/Improving%20Design%20Processes%20through%20better%20Decisio nMaking%20%20an%20experiment%20with%20a%20decision%20making%20supp ort%20tool.pdf

Archer, L. B. (1965). Systematic method for designers. In N. Cross (Ed.), *Developments in Design Methodology* (pp. 57–82). Wiley.

Aristotle (1954). *Rhetoric and Poetics* (W. E. Roberts, Trans.). Random House. (Original work published circa 330 BCE).

Arnold, M. B. and Gasson, J. A. (1954). *The Human Person: An Approach to an Integral Theory of Personality*. Ronald Press.

Arnold, M. B. (1960). *Emotion and Personality: Vol. 1. Psychological Aspects*. Colombia University Press.

Austin, R. and Devlin, L. (2003). *Artful Making: What Managers Need to Know About How Artists Work*. Pearson Education Inc.

Averill, J. R. (1975). A semantic atlas of emotional concepts. *JSAS Catalog of Selected Documents in Psychology*, *5*, 330–421.

Baudrillard, J. (1981). *For a Critique of the Political Economy of the Sign*. Telos Press.

Bechara, A. and Damasio, A. (2005). The somatic marker hypothesis: A neural theory of economic decision-making. *Game and Economic Behavior*, *52*, 336–337.

Ben-Peshat, M. (2004). Popular design and cultural identities – Emotional exchange: Study cases in Israel. *Proceedings of the 4th International Conference on Design and Emotion*. Retrieved 26 January 2011, from www.designandemotion.org

Best, K. (2006). *Design Management: Managing Design Strategy, Process and Implementation*. AVA Publishing.

Biggs, J. B. and Watkins, D. (2001). The paradox of the Chinese learner and beyond. In D. Watkins and J. B. Biggs (Eds), *Teaching the Chinese Learner: Psychological and Pedagogical Perspectives* (pp. 3–23). Comparative Education Research Centre, The University of Hong Kong and The Australian Council for Educational Research Ltd.

Bowes, B. (1999). *A Colour Atlas of Plant Propagation and Conservation*. CRC Press.

Bradley, M. M., Codispoti, M., Sabatinelli, D., and Lang, P. J. (2001). Emotion and motivation II: Sex differences in picture processing. *Emotion*, *1*(3), 300–319.

Brissaud, D., Garro, O., and Poveda, O. (2003). Design process rationale capture and support by abstraction of criteria. *Research in Engineering Design*, *14*, 162–172. Retrieved from EBSCOhost.

Buck, R., Miller, R. E., and Caul, W. F. (1974). Sex, personality, and psychological variables in the communication of affect via facial expression. *Journal of Personality and Social Psychology*, *30*, 587–596.

Cacioppo, J. T., Berntson, G. G., Larsen, J. T., Poehlmann, K. M., and Ito, T. A. (2001). The psychophysiology of emotion. In M. Lewis and J. M. Haviland-Jones (Eds), *Handbook of Emotions* (pp. 173–191). (2nd ed.). Guilford Press.

Camus, A. (1996). *The First Man*. Knopf.

Carlson, R. (1997). *Experienced Cognition*. Lawrence Erlbaum.

Chattaraman, V., Sankar, C. S., and Vallone, A. (2010). Action learning: Application to case study development in graduate education design. *Art, Design, and Communication in Higher Education*, *9*(2), 183–198.

Chau, C. (2007). *The Non-stop Game: Carrie Chau's Drawing Book*. Homeless Ltd.

Chen, A. and Chen, R. (2004). An adaptive design process generated by the integration of systematic design process and design patent protection mechanism. *International Journal of General Systems*, *33*(6), 635–653. Retrieved from Arts & Humanities Citation Index.

Chhibber, S., Porter, C. S., Porter, J. M., and Healey, L. (2004). Designing pleasure; designers' needs. *Proceedings of the 5th International Conference on Design and Emotion*. Retrieved 26 January 2011, from www.designandemotion.org

Chitturi, R. (2009). Emotions by design: A consumer perspective. *International Journal of Design*, *3*(2), 7–17.

Choi, S. (2006). Emotional universal design – Beyond usability of products. *Proceedings of the 5th International Conference on Design and Emotion*. Retrieved 26 January 2011, from www.designandemotion.org

Clarke, S. and Roome, N. (1999). Sustainable business: Learning – action networks as organizational assets. *Business Strategy and the Environment*, *8*, 296–310.

Clore, G. L. (1994). Why emotions vary in intensity. In P. Ekman and R. J. Davidson (Eds), *The Nature of Emotion: Fundamental Questions* (pp. 386–393). Oxford University Press.

Cooper, A. (1999). *The Inmates Are Running the Asylum*. Sams-Pearson Education.

Creswell, J. W. (2008). *Educational Research: Planning, Conducting, and Evaluating Quantitative and Qualitative Research*. (3rd ed.). Pearson/Merrill Prentice Hall.

Creusen, M. E. H. (1998). *Product Appearance and Consumer Choice*. Delft University of Technology.

Cross, M. and Sivaloganathan, S. (2004). A methodology for developing company-specific design process models. *Proceedings of the Institution of Mechanical Engineers: Journal of Engineering Manufacture*, *219* Part B, 265–282. Retrieved from HW Wilson Web.

Csikszentmihalyi, M. (1996). *Flow and the Psychology of Discovery and Invention*. Harper Collins.

Csikszentmihalyi, M. and Rochberg-Halton, E. (1981). *The Meaning of Things: Domestic Symbols and the Self*. Cambridge University Press.

Cupchik, G. C. (2004). The design of emotion. In D. McDonagh, P. Hekkert, and J. van Erp (Eds), *Design and Emotion: The Experience of Everyday Things* (pp. 3–6). Taylor & Francis.

Damasio, A. R. (1994). *Descartes' Error: Emotion, Reason and the Human Brain*. Putnam and Sons.

Darwin, C. (2007). *The Expression of the Emotions in Man and Animals*. Murray.

Davidson, R. J. and Cacioppo, J. T. (1992). New developments in the scientific study of emotion, *Psychological Science*, *3*, 21–22.

Demirbilek, O. and Sener, B. (2002). Emotionally rich products: The effect of childhood heroes, comics and cartoon characters. In D. McDonagh, P. Hekkert, and J. van Erp (Eds), *Design and Emotion: The Experience of Everyday Things* (pp. 278–283). Taylor & Francis.

Denton, H. G., McDonagh, D., Baker, S., and Wormald, P. (2004). Introducing the student designer to the role of emotion in design. In McDonagh, P. Hekkert, and J. van Erp (Eds), *Design and Emotion: The Experience of everyday Things* (pp. 415–420). Taylor & Francis.

Desmet, P. M. A. (1999). To love and not to love: Why do products elicit mixed emotions? *Proceedings of the 1st International Conference Design and Emotion.* Retrieved 26 January 2011, from www.designandemotion.org

Desmet, P. M. A. (2002). *Designing Emotions.* Delft University of Technology.

Desmet, P. M. A. (2003). A multilayered model of product emotions. *The Design Journal, 6*(2), 4–13.

Desmet, P. M. A. (2008). Inspire and desire. In P. M. A. Desmet, J. van Erp, and M. A. Karlsson (Eds), *Design and Emotion Moves* (pp. 108–127). Cambridge Scholars Publishing.

Desmet, P. M. A. and Hekkert, P. (2002). The basis of product emotions. In W. S. Green and P. W. Jordan (Eds), *Pleasure with Products, Beyond Usability* (pp. 61–67). Taylor & Francis.

Desmet, P. M. A. and Hekkert, P. (2009). Special issue editorial: Design & emotion. *International Journal of Design, 3*(2), 1–6.

Dewey, J. (1934). *Art as Experience.* Penguin Putnam.

Dunne, A. (1999). *Hertzian Tales Electronic Products, Aesthetic Experience and Critical Design.* RCA Research Publications.

Dweck, C.S., Hing, Y.Y., and Chiu, C.Y. (1993). Implicit theories: individual differences in likelihood and meaning of dispositional inferences. *School Psychology Review, 13*, 162–170.

Enayati, J. (2002). The research: Effective communication and decision-making in diverse groups. In M. Hemmati, F. Dodds, and J. Enayati (Eds), *Multi-Stakeholder Processes for Governance and Sustainability: Beyond Deadlock and Conflict* (pp. 73–95). Earthscan Publications.

Ekman, P. (1972). Expression and the nature of emotion. In K. R. Scherer and P. Ekman (Eds), *Approaches to Emotion* (pp. 319–344). Lawrence Erlbaum.

Ekman, P. and Friesen, W. V. (1975). *Unmasking the Face: A Guide to Recognizing Emotions from Facial Cues.* Prentice-Hall.

Enders, G. (2004). Design practice presentation on tools and methods for emotion-driven design. *Proceedings of the 4th International Conference on Design and Emotion.* Retrieved 26 January 2011, from www.designandemotion.org

Erev, I. and Haruvy, E. (2013). Learning and the economics of small decisions. *The Handbook of Experimental Economics, 2*, 638–700.

Evans, M. and Sommerville, S. (2007). A design for life: Futures thinking in the design curriculum. *Futures Research Quarterly, 23*(3), 5.

Flam, H. (1993). Fear, loyalty and greedy organizations. In S. Fineman (Ed.), *Emotion in Organizations* (pp. 58–75). Sage.

Flanagan, J. S. (1954). The critical incident technique. *Psychological Bulletin, 51*, 327–358.

Foa, E. B. and Kozak, M. J. (1986). Emotional processing of fear: exposure to corrective information. *Psychological Bulletin, 99*(1), 20.

Foque, R. (1995). Designing for patients: A strategy for introducing human scale in hospital design. *Design Studies, 16*, 29–49.

Forlizzi, J., Disalvo, C., and Hannington, B. (2003). On the relationship between emotion, experience and the design of new products. *The Design Journal*, *6*(2), 29–38.

Fraenkel, J. R. and Wallen, N. E. (2008). *How to Design and Evaluate Research in Education*. McGraw Hill Higher Education.

Fredrickson, B. L. (2000). Cultivating positive emotions to optimize health and well-being. *Prevention and Treatment*, 3, Article 1. http://journals.apa.org/prevention/volume3/pre0 030001a.html

Freire, P. (1974). Conscientisation. *CrossCurrents*, *24*(1), 23–31.

Frijda, N. H. (1986). *The Emotions*. Cambridge University Press.

Fung, A., Lo, A., and Rao, N. M. (2005). *Creative Tools*. School of Design, Hong Kong Polytechnic University.

Funke, R. (1999). Emotions: The key to motivation in the experience society. *Proceedings of the 1st International Conference on Design and Emotion*. Retrieved 26 January 2011, from www.designandemotion.org

Gage, N. L. and Berliner, D. C. (1992). *Educational Psychology*. Houghton Mifflin.

Gaver, W. W. (1999). Irrational aspects of technology: Anecdotal evidence. In C. J. Overbeeke and P. Hekkert (Eds), *Proceedings of the 1st International Conference on Design and Emotion* (pp. 47–54). Delft University of Technology.

Gayretli, A. and Abdalla, H. S. (1999). A prototype constraint-based system for the automation and optimization of machining processes. *Proceedings of the Institution of Mechanical Engineers: Journal of Engineering Manufacture*, *213* Part B. Retrieved from Arts & Humanities Citation Index.

Gibbons, A. (2003, 10 Jan.). Multiculturalism just isn't enough. *Times Educational Supplement*.

Gladstones, W. A. H. (1962). A multidimensional study of facial expression of emotion. *Australian Journal of Psychology*, *14*, 95–99.

Golec de Zavala, A., Keenan, O., Ziegler, M., Ciesielski, P., Wahl, J. E., and Mazurkiewicz, M. (2023). App-based mindfulness training supported eudaimonic wellbeing during the COVID19 pandemic. *Applied Psychology: Health and Well-Being*, 1–18.

Goldschmidt, G. (1999). Design. In M. A. Runco and S. R. Pritzker (Eds), *Encyclopedia of Creativity* (p. 525). Academic Press.

Goleman, D. (2004). *Emotional Intelligence & Working with Emotional Intelligence*. Bloomsbury.

Govers, P., Hekkert, P., and Schoormans, J. P. L. (2004). Happy, cute and tough: Can designers create a product personality that consumers understand? In D. McDonagh, P. Hekkert and J. van Erp (Eds), *Design and Emotion: The Experience of Everyday Things* (pp. 345–350). Taylor & Francis.

Green, P. (1974). *Design Education*. Batsford.

Guba, E. and Lincoln, Y. (1981). *Effective Evaluation: Improving the Usefulness of Evaluation Results Through Responsiveness and Naturalistic Approaches*. Jossey-Bass.

Hall, J. A., Carter, J. D., and Horgan, T. G. (2000). Gender differences in nonverbal communication of emotion. In A. Fischer (Ed.), *Gender and Emotion: Social Psychological Perspectives* (pp. 97–117). Cambridge University Press.

Hall T., Strangman, N., and Meyer, A. (2009). *Differentiated Instruction and Implications for UDL Implementation*. National Center for Accessing the General Curriculum (USA).

Hakatie, A. and Ryynänen, T. (2006). Product attributes and the model of emotional design: How do the product development engineers perceive product features? *Proceedings of the 5th International Conference on Design and Emotion*. Retrieved 26 January 2011, from www.designandemotion.org

Harada, K. (1999). Performance based codes and performance based fire safety design. *Fire Science and Technology*, *19*(1), 1–10.

Herrmann, J. W. and Schmidt, L. C. (2002). Viewing product development as a decision production system. *Proceedings of Design Engineering Technical Conferences and Computers and Information in Engineering Conference*. Montreal, Canada, August 2002.

Heskett, J. (Ed.) (2007). *Very Hong Kong: Design 1997–2007*. Hong Kong Design Centre.

Hillier, B., Musgrove, J., and O' Sullivan, P. (1984). Knowledge and design. In N. Cross (Ed.), *Developments in Design Methodology* (pp. 245–264). Wiley.

Ho, A. G. and Siu, K. W. M. (2009a). Emotionalise design, emotional design, emotion design. *Proceedings of the International Association of Societies of Design Research*. Retrieved 25 January 2011, from www.iasdr2009.org/ap/Papers/Orally%20Presented%20Papers/Behavior/Emotionalise%20Design,%20Emotional%20Design,%20Emotion%20Des ign%20-%20A%20new%20perspective%20to%20understand%20their%20relationsh ips.pdf/

Ho, A. G. and Siu, K. W. M. (2009b). Role of designers in the new perspective of design and emotion. *Design Principles and Practices: An International Journal, 4*(3), 15–24.

Ho, A. G. (2010). Exploring the relationships between emotion and design process for designers today. *Proceedings of the 1st International Conference on Design Creativity ICDC 2010*. The Design Society (Japan).

Ho, A. G. and Siu, K. W. M. (2012). Emotion design, emotional design, emotionalised design: A review on their relationships from a new perspective. *The Design Journal, 15*(1): 9–32.

Ho, A. G. (2014). The new relationship between emotion and the design process for designers. *Archives of Design Research, 27*(2), 45–55.

Holper, C. V. (1998). Affective education in French-speaking parts of Belgium: A brief overview. In Y. J. Katz, P. Lang, and I. Menezes (Eds), *Affective Education: A Comparative View* (pp. 19–27). Cassell.

Horton, L. (1971). Teacher education: By design or crisis? *Journal of Teacher Education, 22*(3), 265–267.

Hummels, C. (1999). Engaging contexts to evoke experiences. In C. J. Overbeeke and P. Hekkert (Eds), *Proceedings of the 1st International Conference on Design and Emotion* (pp. 39–46). Delft University of Technology.

Icons-Land (2006). *Vista Style Emoticons*, Icons-Land. Retrieved 26 January 2011, from www.icons-land.com/productvistastyleemoticons.php

Jardón, C. M. F. (2011). Deployment of Core Competencies to obtain success in SMEs. *Depatamento de Economia Aplicada, 1–27*.

James, W. (2003). What is an emotion? In R. C. Solomon (Ed.), *What is an Emotion?: Classic and Contemporary Readings* (pp. 66–77). Oxford University Press.

Jensen, R. (1999). *The Dream Society: The Coming Shift from Information to Imagination*. McGraw-Hill.

Johnson, D. W. and Johnson R. (2005). Classroom conflict: Controversy over debate in learning groups. *American Journal of Educational Research, 22*, 237–256.

Jon, S. and Greene, R. W. (2003). *Sociology and You*. Glencoe McGraw-Hill.

Jones, J. C. (1984). A method of systematic design. In N. Cross (Ed.), *Developments in Design Methodology* (pp. 9–31). Wiley.

Jordan, P. (2000). *Designing Pleasurable Products: An Introduction to the New Human Factors*. Taylor & Francis.

Kaufmann, G. (2003). The effect of mood on creativity in the innovative process. In L. V. Shavinina (Ed.), *The International Handbook on Innovation* (pp. 191–203). Elsevier.

Karl, A. (2006). *The Design Process*. Fairchild.

Keedy, J. (1998). Graphic design in the postmodern era. *Emigre 47 Summer*, 50–60.

Kondoyianni, A., Short, R.K., and Sideri, A. (1998). Affective education and the green school curriculum. In Y. J. Katz, P. Lang, and I. Menezes (Eds), *Affective Education: A Comparative View* (pp. 58–71). Cassell.

Koskinen, I., Battarbee, K., and Mattelmäki, T. (Eds) (2003). *Emphatic Design: User Experience in Product Design*. IT Press.

Kramer, R. (2008). Learning how to learn: Action learning for leadership development. *Innovations in Public Leadership Development*, 296–326.

Lacey, E. (2009). Contemporary ceramic design for meaningful interaction and emotional durability: A case study. *International Journal of Design*, *3*(2), 87–92.

Lang, P. J., Bradley, M. M., and Cuthbert, B. N. (1998). Emotion, motivation, and anxiety: Brain mechanisms and psychophysiology. *Biological Psychiatry*, *44*(12), 1248–1263.

Lawson, B. (2006). *How Designers Think: The Design Process Demystified* (4th ed.). Elsevier/ Architectural Press.

Lazarus, R. S., Kanner, A. D., and Folkman, S. (1980). Emotions: A cognitive-phenomenological analysis. In R. Plutchik and H. Kellerman (Eds), *Theories of Emotion, vol. 1: Emotion: Theory, Research, and Experience* (pp. 189–217). Academic Press.

Lazarus, R. S. (1993). From psychological stress to the emotions: A history of changing outlooks. *Annual Review of Psychology*, *44*, 1–21.

Levin, P. H. (1984). Decision-making in urban design. In N. Cross (Ed.), *Developments in Design Methodology* (pp. 107–122). Wiley.

Li, L. and Barrett, J. (2002). Action learning research in engineering design teaching. *Action Learning Research in Engineering Design Teaching*, pp. 391–401.

Lo, K. P. Y. (2007). Emotional design for hotel stay experiences: Research on guest emotions and design opportunities. *International Association of 50 Societies of Design Research* [Press release]. Retrieved from International Association of 50 Societies of Design Research Ondisc Database.

Lodico, M. G., Spaulding, D. T., and Voegtle, K. H. (2010). *Methods in Educational Research: From Theory to Practice*. Jossey-Bass.

Longueville, B., Le Cardinal, J., Bocquet, J., and Daneau, P. (2003). Towards a project memory for innovative product design: A decision-making process model. *International Conference on Engineering Design (ICED03)*, Stockholm Publishing.

Lum, B. J. (1997). Student mentality: Intentionalist perspectives about the principal. *Journal of Educational Administration*, *35*(3), 210–233.

Luckman, J. (1967). An approach to the management of design. *Operational Research Quarterly*, *18*, 345–358.

Mace, E. C., Neef, N. A., Shade, D., and Mauro, B. C. (1996). Effects of problem difficulty and reinforcer quality on time allocated to concurrent arithmetic problems. *Journal of Applied Behavior Analysis*, *29*(1), 11–24.

Mann, T. (2004). *Time Management for Architects and Designers: Challenges and Remedies*. W. W. Norton.

Marshall, L. and Rowland, F. (1993). *A Guide to Learning Independently*. Open University Press.

McClenaghan, P. (2007). *Brandscape Architecture: Towards a Unified Theory of Experiential Design: International Association of 50 Societies of Design Research* [Press release]. Retrieved from International Association of 50 Societies of Design Research Ondisc database.

McLennan, J. F. (2004). *The Philosophy of Sustainable Design*. Ecotone Publishing.

Mehrabian, A. and Russell, J. A. (1977). Evidence for a three-factor theory of emotions. *Journal of Research in Personality*, *11*, 273–294.

Miles, M. B. and Huberman, A. M. (1984). *Qualitative data analysis*. Sage.

Mollerup, P. (2005). *Wayshowing: A Guide to Environmental Signage Principles & Practices.* Lars Müller Publishers.

MSN (n.d.). *Use Emoticons in Messages*, Microsoft Corporation. Retrieved 26 January 2011, from http://messenger.msn.com/Resource/Emoticons.aspx

Murty, P. and Purcell, T. (2007). Designerly, reflective and insightful ways of design. *International Association of 50 Societies of Design Research* [Press release]. Retrieved from International Association of 50 Societies of Design Research Ondisc Database.

Naqvi, N., Shiv, B., and Bechara, A. (2006). The role of emotion in decision making: A cognitive neuroscience perspective. *Association for Psychological Science, 15*, 5.

Noble, I. and Bestley, R. (2005). *Visual Research.* AVA Publishing.

Norman, D. A. (2004). *Emotional Design: Why We Love (Or Hate) Everyday Things.* Basic Books.

Ortony, A., Clore, G. L., and Collins, A. (1988). *The Cognitive Structure of Emotions.* Cambridge University Press.

Overbeeke, C. J., and Hekkert, P. (Eds) (1999). *Proceedings of the 1st International Conference on Design and Emotion.* Retrieved 26 January 2011, from www.designandemotion.org

Pahl, G. and Beitz, W. (1996). *Engineering Design* (2nd ed.). Springer.

Pampliega, A.M. and Marroquin, M. (1998). Affective education and the new Spanish educational reforms. In Y. J. Katz, P. Lang, and I. Menezes (Eds), *Affective Education: A Comparative View* (pp. 85–98). Cassell.

Park, T., Lee, J. H., Kim, S., Kim, M., Shim, M., Park, J. J., and Noh, H. (2007). Design process improvement for effective rich interaction design. *International Association of 50 Societies of Design Research* [Press release]. Retrieved from International Association of 50 Societies of Design Research Ondisc Database.

Paterson, M. C. (1979). *Environmental and Genetic Interactions in Human Cancer* (No. AECL–6958).

Peto, J. (Ed.). (1999). *Design Process, Progress, Practice.* Design Museum.

Pekrun, R. (2006). The control-value theory of achievement emotions: Assumptions, corollaries, and implications for educational research and practice. *Educational Psychology Review, 18*, 315–341.

Picard, R. W. (1997). *Affective Computing.* MIT Press.

Pine, J. and Gilmore, J. (1999). *The Experience Economy.* Harvard Business School Press.

Plato (1955). The symposium. In W. Hamilton (Ed.), *The Symposium*, Penguin. (Original work published circa 390 BCE.)

Plutchik, R. (1980). *Emotion: A Psychoevolutionary Synthesis.* Harper & Row.

Reyman, I. M. M. J., Hammer, D. K., Kroes, P. A., van Aken, J. E., Dorst, C. H., Bax, M. F. T., and Basten, T. (2006). A domain-independent descriptive design model and its application to structured reflection on design processes. *Research in Engineering Design 16*, 147–173. Retrieved from Arts & Humanities Citation Index.

Roberts, P. and Burgess, P. (1973). Early work in design. In B. Aylward (Ed.), *Design Education in Schools* (pp. 28–55). Evans.

Rollestone, G. (2003). *Scenario-based, Value-driven Design Methods* [White paper]. Icon Medialab.

Rosella, F. (2002). F+R hugs: How to communicate physical and emotional closeness to a distant loved one? In A. Kurtgozu (Ed.), *Proceedings of the 4th International Conference on Design and Emotion* [CD ROM]. METU Press.

Russell, J. A. (1980). A circumflex model of affect. *Journal of Personality and Social Psychology, 39*, 1161–1178.

Salili, F. and Hoosain, R. (Eds). (2001). *Multicultural Education: Issues, Policies, and Practices.* Information Age Publishing.

Salovey, P. and Mayer, J. D. (1990). Emotional intelligence. Imagination, *Cognition and Personality, 9*(3), 185–211.

Salovey, P. and Lopes, P. N. (2004). Toward a broader education: Social, emotional, and practical skills. In J. E. Zins, R. P. Weissberg, M. C. Wang, and H. J. Walberg (Eds), *Building Academic Success on Social and Emotional Learning: What Does the Research Say* (pp. 76–93). Teachers College Press.

Sanders, L. (1999). Design for experiencing: New tools. In C. J. Overbeeke and P. Hekkert (Eds), *Proceedings of the 1st International Conference on Design and Emotion* (pp. 87–92). Delft University of Technology.

Schachter, S. and Singer, J. (1962). Cognitive, social, and physiological determinants of emotional state. *Psychological Review, 69*(5), 379.

Scaletsky, C. C. and Marques, A. C. (2009). Materials database organization for the design process. *Proceedings of the 8th European Academy of Design Conference* (pp. 421–425). Robert Gordon University.

Scherer, K. R. (1984). On the nature and function of emotion: A component process approach. In K. R. Scherer and P. Ekman (Eds), *Approaches to Emotion* (pp. 293–317). Lawrence Erlbaum.

Scherer, K. R. and Tran, V. (2001). Effects of emotion on the process of organisational learning. In A. Berthoin Antal, J. Child, M. Dierkes, and I. Nonaka (Eds), *Handbook of Organizational Learning and Knowledge* (pp. 369–392). Oxford University Press.

Schimmack, U., Oishi, S., and Diener, E. (2002). Cultural influences on the relation between pleasant and unpleasant emotions: Asian dialectic philosophies or individualism-collectivism. *Cognition and Emotion, 16*, 705–719.

Schmitt, B. H. (1999). *Experiential Marketing*. Free Press.

Schweitzer, M. E. and Cachon, G. P. (2000). Decision bias in the newsvendor problem with a known demand distribution: Experimental evidence. *Management Science, 46*(3), 404–420.

Seidlitz, L. and Diener, E. (1998). Gender differences in the recall of affective experiences. *Journal of Personality and Social Psychology, 74*, 262–271.

Seymour, R. and Powell, R. (2003). Emotional ergonomics. *Design Week, 18*(50), 5.

Skinner, B. F. (1968). *The Technology of Teaching*. Appleton-Century-Crofts.

Siu, K. W. M. (2003). Creating potential problems: Knowledge, experience and critical thinking. *6th Asian Design Conference* (Presentation).

Somekh, B. and Noffke, S. E. (2009). *The SAGE Handbook of Educational Action Research*. SAGE, pp. 1–568.

Spendlove, D. (2007). A conceptualisation of emotion within art and design education: A creative, learning and product orientated triadic schema. *International Journal of Art and Design Education, 26*(2), 155–166.

Spillers, F. (2004). Emotion as a cognitive artifact and the design implications for products that are perceived as pleasurable. *Proceedings of the 4th International Conference on Design and Emotion*. Retrieved 26 January 2011, from www.designandemotion.org

Sternberg, R. J. (Ed.). (1999). *Handbook of Creativity*. Cambridge University Press.

Stevenson, D. (2019). Branding, promotion, and the tourist city. In *The Routledge Companion to Urban Media and Communication* (pp. 265–273). Routledge.

Stumpf, S. (2001). *Analysis and Representation of Rhetorical Construction of Understanding in Design Teams' Experiential Learning*. British Thesis Service, British Library.

Suri, J. F. (2003). The experience of evolution: Developments in design practice. *The Design Journal, 6*(2), 39–48.

Sutherland, S. R. (2002). *Higher Education in Hong Kong: Report of the University Grants Committee*. Secretary for Education and Manpower. Retrieved 26 January 2011, from www.e-c.edu.hk/eng/reform/report/forward.html

Tan, E. (1999). Emotion and style as design. In C. J. Overbeeke and P. Hekkert (Eds), *Proceedings of the 1st International Conference on Design and Emotion* (pp. 55–64). Delft University of Technology.

Takashi, A., Jumpei, K., Yoshiki U., and Yoshiyuki, M., (2007). Classification and guidelines for selection of design modelling methods. *International Association of 50 Societies of Design Research* [Press release]. Retrieved from International Association of 50 Societies of Design Research Ondisc Database.

Tassinary, L. G. and Cacioppo, J. T. (1992). Unobservable facial actions and emotion. *Psychological Science, 3,* 28–33.

Thackara, J. (1997). *Winners: How Successful Companies Innovate by Design.* Gower Publishing.

Ting, C. C. (2007). A study on the determinants of form in digital camera as they affect usability. *International Association of 50 Societies of Design Research* [Press release]. Retrieved from International Association of 50 Societies of Design Research Ondisc Database.

Tomkin, S. S. (1995). Evolution of the affect system. In S. Tomkin and E. V. Demos (Eds), *Exploring Affect: The Selected Writings of Sylvan* (pp. 66–77). Cambridge University Press.

Tomkin, S. S. (2008). *Affect Imagery Consciousness.* Springer Publishing.

Turner, J. E. and Schallert, D. L. (2001). Expectancy–value relationships of shame reactions and shame resiliency. *Journal of Educational Psychology,* 93(2), 320.

Tzvetanova, S. (2007). Emotional interface methodology. *International Association of 50 Societies of Design Research.* Retrieved 26 January 2011, from International Association of 50 Societies of Design Research Ondisc Database.

University Grants Committee [UGC]. (2001). *Document for the Open Forum on Higher Education in Hong Kong,* 23 October 2001, Hong Kong Polytechnic University.

van Aken, J. E. (2005). Valid knowledge for the professional design of large and complex design processes. *Design Studies,* 26(4), 379–404. Retrieved from Design and Applied Arts Index.

Vince, R. and Martin, L. (1993). *Management Education and Development,* 24(3), 205–215.

Visocky O'Grady, J. and Visocky O'Grady, K. (2017). *A Designer's Research Manual, Updated and Expanded: Succeed in Design by Knowing Your Clients and Understanding What They Really Need.* Rockport Publishers.

Vosburg, S. K. (1998). The effects of positive and negative mood on divergent-thinking performance. *Creativity Research Journal, 11*(2), 165–172.

Vygotsky, L. S. (1978). *Mind in Society: The Development of Higher Psychological Processes.* Massachusetts Institute of Technology Press.

Wagner, H. L., Buck, R., and Winterbotham, M. (1993). Communication of specific emotions: Gender differences in sending accuracy and communication measures. *Journal of Nonverbal Behavior, 17,* 29–53.

Walker, J. A. (1989). *Design History and the History of Design.* Billing & Sons.

Wallen, N. E. and Fraenkel, J. R. (1991). *Educational Research: A Guide to the Process.* Routledge. https://doi.org/10.4324/9781410601001

Wickens, C. D. and Hollands, J. G. (2000). *Engineering Psychology and Human Performance.* Prentice Hall.

Wogalter, M. S., Dejoy, D. M., and Laughery, K. R. (1999). *Warnings and Risk Communication.* Taylor & Francis.

Wundt, W. (1905). *Fundamentals of Psychology* (7th ed.). Engelman.

Yeomans, M. (1990). The future of design in further and higher education. In D. Thistlewood (Ed.) *Issues in Design Education* (pp. 166–184). Longman.

Zabrondin, Y., Popova, M., and Minaev, A. (1998). Affective education: A Russian perspective. In Y. J. Katz, P. Lang, and I. Menezes (Eds), *Affective Education: A Comparative View* (pp. 46–57). Cassell.

Zeidner, M. (1998). *Test Anxiety: The State of the Art*. Springer.

Zeidner, M. (2007). Test anxiety in educational contexts: Concepts, findings, and future directions. In *Emotion in Education* (pp. 165–184). Academic Press.

Zimmerman, B. J. (2001). Theories of self-regulated learning and academic achievement. In B. J. Zimmerman and D. H. Schunk (Eds), *Self-regulated Learning and Academic Achievement* (pp. 1–36). Lawrence Erlbaum Associates.

Index

For Product Safety Concerns and Information please contact our EU
representative GPSR@taylorandfrancis.com
Taylor & Francis Verlag GmbH, Kaufingerstraße 24, 80331 München, Germany